THE SOCIAL CONTEXT OF TECHNOLOGICAL EXPERIENCES

This book demonstrates how technology and society shape one another, and that there are intrinsic connections between technological experiences and social relationships. It employs an array of theoretical concepts and methodological tools to examine the technology – society nexus among three urban groups in India (traditional caste-based handloom weavers, subaltern Dalit communities, and informal female labour).

It provides evidence of how innovations such as industrial technologies, communication technologies, and workplace technologies are not only about strides in science and engineering but also about politics and sociology on the ground. The book contributes to the growing research in innovation studies and technology policy that establishes how technological processes and outcomes are contingent on complex sociological variables and contexts. The author offers an inclusive, holistic, and interdisciplinary approach to understanding the field of innovation and technological change and development by involving various methodologies (network analysis, archival work, oral histories, focus group discussions, interviews).

The book will serve as reference for researchers and scholars in social sciences, especially those interested in development studies, science and technology policy and innovation studies, information and communication technology (ICT) policy, public policy, management, social work and research methods, economics, sociology, social exclusion and subaltern studies, women's studies, and South Asian studies. It will also be useful to nongovernmental organisations, activists, and policymakers.

Anant Kamath is a social scientist based in Bangalore, India. He has taught development, social research, and technological change at Azim Premji University, Bangalore. Previously, he was a scholar at the United Nations University – Maastricht Economic and Social Research Institute on Innovation and Technology (UNU-MERIT) in The Netherlands, the Centre for Development Studies (CDS) in Thiruvananthapuram, and the Madras School of Economics. His research interests are in the economic sociology of technological change and experiences, and in the political economy of development. He is also involved in the western classical music scene in Bangalore, as principal violinist and lead of the BSM Chamber Orchestra.

SCIENCE AND TECHNOLOGY STUDIES
Series Editor: Sundar Sarukkai
*Former Professor of Philosophy, National Institute of
Advanced Studies, Bengaluru, India, and Founder-Director,
Manipal Centre for Philosophy and Humanities*

There is little doubt that science and technology are the most influential agents of global circulation of cultures. Science & Technology Studies (STS) is a well-established discipline that has for some time challenged simplistic understanding of science and technology (S&T) by drawing on perspectives from history, philosophy and sociology. However, an asymmetry between 'western' and 'eastern' cultures continues, not only in the production of new S&T but also in their analysis. At the same time, these cultures which have little contribution to the understanding of S&T are also becoming their dominant consumers. More importantly, S&T are themselves getting modified through the interaction with the historical, cultural and philosophical worldviews of the non-western cultures and this is creating new spaces for the interpretation and application of S&T. This series takes into account these perspectives and sets right this global imbalance by promoting monographs and edited volumes which analyse S&T from multicultural and comparative perspectives.

For more information about this series, please visit: www.routledge.com/Science-and-Technology-Studies/book-series/STS

THE SOCIAL CONTEXT OF TECHNOLOGICAL EXPERIENCES

Three Studies from India

Anant Kamath

Routledge
Taylor & Francis Group

LONDON AND NEW YORK

First published 2020
by Routledge
2 Park Square, Milton Park, Abingdon, Oxon OX14 4RN

and by Routledge
52 Vanderbilt Avenue, New York, NY 10017

Routledge is an imprint of the Taylor & Francis Group, an informa business

© 2020 Anant Kamath

The right of Anant Kamath to be identified as author of this work has been asserted by him in accordance with sections 77 and 78 of the Copyright, Designs and Patents Act 1988.

British Library Cataloguing-in-Publication Data
A catalogue record for this book is available from the British Library

Library of Congress Cataloging-in-Publication Data
A catalog record for this book has been requested

ISBN: 978-1-138-32408-4 (hbk)
ISBN: 978-0-367-49506-0 (pbk)
ISBN: 978-1-003-03264-9 (ebk)

Typeset in Goudy
by Apex CoVantage, LLC

CONTENTS

ILLUSTRATIONS

Figures

Tables

ACKNOWLEDGEMENTS

The first case study on weavers in southern Kerala was undertaken at UNU-MERIT (United Nations University – Maastricht Economic and Social Research Institute on Innovation and Technology), Maastricht, The Netherlands, with its financial support. Here, I would like to thank Robin Cowan, Sunil Mani, Saurabh Arora, Adam Szirmai, Ad Notten, and Eveline in de Braek. A good part of the study was undertaken at the Centre for Development Studies (CDS), Thiruvananthapuram, India, an institution to which I am always indebted. Two libraries, besides the UNU-MERIT/Maastricht University libraries, were greatly exploited for bibliographic material for this study. These are the KN Raj Library at CDS, and the University of Georgia Library at UGA in the United States. To these libraries, and their staff, I owe deep gratitude. For empirical work, I thank J Devika (CDS), Mohanakumar, Ramesan, Viswambaran, Sudhakaran at Balaramapuram; and to PS Mani. Special thanks are due to the Saliyar community at Balaramapuram, particularly Selvaraj, Subramanian and family, and Magesh and family, Paramasivan and Manikandan; and to the handloom weaving community at Payattuvila, particularly Udayan and Thankappan Panicker, and family.

I thank Azim Premji University, Bangalore, for financial support for the second case study on the Dalit communities in peri-urban Bangalore. I must particularly mention the excellent research assistance of Vinay Kumar. Acknowledgements are due to Balmurli Natrajan, Aparna Sundar, Gayatri Menon, and Anil Sethi for early discussions; Sundar Sarukkai, Aseem Prakash, Satish Deshpande, Rakesh Basant, Bala Subrahmanya, and Nandini Chami for discussions on improving the analysis; Cynthia Stephen, Raja Nayak, the many Dalit political and civil society organisations I interviewed, the Dalit Indian Chamber of Commerce and Industry, and families at Choodasandra.

For the third case study, my coauthor Neethi and I must mention the extraordinary perseverance of our third coauthor, Saloni Mundra, who undertook a great part of the empirical work as our postgraduate student. Acknowledgements are due to the several pourakarmikas spoken to, the BBMP Ward Office of Koramangala Block 5, and Lekha and the rest of the staff at Alternative Law Form.

Due thanks to Raji, who facilitated the uninterrupted flow of writing. My family deserves utmost acknowledgement for moral support throughout. While my father will not be able to see this in its final form, our dearest daughter Madhavi surely will hold this book in her tiny hands, and will hopefully forgive me for spending considerable lengths of time away from her, completing this work day after day. And of course, to Neethi, firstly and finally, the axis of life and work, and my lifelong partner in Tennysonian efforts in striving, seeking, finding, without yielding.

1

INTRODUCTION

From stately temples of modernity to sleek silver bullets

At first, it appears amusing. People hurling floral offerings and salutations to a monumental dam with all sluices open and torrents of water thundering down its slopes; rural folk paying obeisance to a great technological feat by India, for India. A dramatised setting, of course. Nevertheless, this short film of under a dozen minutes, *Symphony of Progress*, created in 1972 by the state-run Films Division, would have stirred nothing short of sheer exhilaration within its audience at the time, celebrating not only the country's economic progress but also, importantly, adulating the central role of big technology in this titanic endeavour. The then prime minister, Jawaharlal Nehru, eloquently paid homage to big dams as "the temples of modern India," consecrating, by extension, big dam technology – or even technological progress as their principal deity. And while this event within the film can evoke both a smirk and a shudder, Nehru's veneration was not unfounded, as he strongly believed that these symphonies of technological progress could bear fruit in the long term only if steeped within the more noble bedrock of scientific temper in Indian economy and society. This idea emits the powerful message that technological outcomes and progress actuate social change. But if we put our ears against the ground and pay close attention to the character of this bedrock, it radiates an unspoken but powerful message that technological outcomes and progress are also unmistakably sculpted in turn by forces deep beneath – namely its society.

The purpose of this book is to thrust forward exactly this perspective – that technological outcomes and processes have sociological underpinnings, and that technology and society mould one another in complex harmony, procreating nonobvious development outcomes. To understand the upshot in one, we have to study the dynamics of the other. On employing this perspective, we embrace a view of the relationship between technological experience and society not as an oversimplified linear cause–effect relationship but as an explicable and decodable consonance of these two restless forces.

A technological experience (an engagement, practice, involvement, formal or informal association, or even the process of gaining familiarity around a technology) is rooted within social structure and social functions, be it around an artefact such as mobile phone, or a process to improve work experiences and vanquish monetary fraud, or even subtleties in knowledge and information around artisanal production practices. The counterpart of this process, i.e., where technology sets in motion social shifts, is more easily comprehensible. On first sight, these relationships might appear plainly obvious. However, this book demonstrates how pertinent an enquiry into this consonance of technology and society actually is (both for our contemporary era as well as our future ahead); how cavernous the gap in policy and popular (possibly even academic) discourse has been on this theme; how multiplicity in concept, method, site, and subject is warranted; and how multilayered and fine-grained socio-technological experiences actually are in lived reality. Bypassing these actualities, either out of escapist convenience or out of ignorance, has ushered us into a comfortable position that developmental challenges, if to be mitigated by new technologies, might face obstruction related to technical or engineering complexities alone. For example: the challenges that the digital divide bring are hard but are popularly believed to be best dealt with when laid upon the operating table of the technologist and state actor, who work together and deliberate with sweat and toil on sophisticated or esoteric technological solutions; however, the complexities of the sociological crucible harbouring the digital divide are glaringly missing on that table and in those deliberations, while they are plainly visible in reality. While the efforts of the aforesaid actors are critical and must certainly not be in vain, they usually are estranged from the acknowledgement, appreciation, and inclusion of the social context from the technological experience.

The task ahead, therefore, is to reconcile this disunity and work on a synthesis in understanding the sociological bases of technological experience and outcome. To commence this synthesis, I invite the readers to wade through the murky waters of technological determinism and to realise the magnitude of its stronghold in indoctrinating notions around the role of technology as the steering wheel of development. I urge the readers to take another look at popular notions of "knowledge economy" and "information society" and the policy formulations that placate these. I also provoke the realisation how expected dividends from our manufactured "contemporary technological era" are nothing but quick-fix solutionist inclinations masquerading through the development dialogue disguised as progressive ideals. I ask to expand the question, "What technologies do people encounter?" to include the angle of "What kind of people do technologies encounter?" And finally, I escort the readers in meandering through the society–technology polyphony in various settings, each one uniquely labyrinthine.

Often, I may present dismal outcomes to explore the relationship between technology and society. However, my aim is not to pass a baton loaded with melancholy or bleakness but actually to deepen perspective and expand into new spectra of observation beyond what is traditionally witnessed (or overlooked) by policy or popular discourse in the technology–society nexus in India. My purpose through this volume is not to espouse a new theoretical treatise or to provide a cultural critique around technology. Instead, I present the socio-technological experience through a wide palette of theoretical concepts (from social capital, to social constructionism, to gendered labour informality) and methodological tools (from quantitative network analysis to micro oral histories) in three urban settings (small towns, the peri-urban, and the metropolitan) among three groups (traditional weavers, subaltern castes, and informal female labour). Importantly, I also appeal to the dire necessity for greater enquiry into the technological engagement of the *subaltern*, in terms of historically deprived or socioeconomically downwardly mobile groups. This is one area of examination in India that is appallingly deficient, and urgently needs greater contemplation in analysis and action in policy and programme.

Overall, I wish to drive home the verity, well established in the literature, that technologies are not solely about strides in science and engineering but also include politics and sociology. As a result of this, an interplay of an array of sociopolitical factors can generate unexpected or often imperceptible technological outcomes across socioeconomic settings, thus dethroning the streamlined, standardised, and overconfident technological prescriptions that have been built on the basis of glowing views of technological processes and on the nature of technology as such. We live, operate, and aspire in a socio-technical system. Hence, the ultimate purpose of the expedition ahead is to gain a grasp on the wondrous yet elusive complexity of the reality we call a socio-technological experience.

Technological determinism and its discontents

What am I presenting the "socio-technological experience" in contrast to? What is the meta-narrative around the role of technology in development in India? How has it been treated in the academic literature? These initial questions set the keystone to begin our journey on the social context of technological experiences in India. To understand what we're up against, we are required to immerse ourselves first in an understanding of what technological determinism and technological solutionism is, based on the rich literature in the domain.

When we imagine for a moment what "technology" means, it immediately conjures up images of sleek-looking and educated technicians pondering over complex instruments or computer monitors (Marx, 1994, 1997, 2010). In fact, a summary test of this over a class of students, in my own experience year after year, brings up the same mental picture – "technology" immediately resonates with visions of high-sophistication digital technologies and a spotless work environment among mostly white-collar, English-speaking male workers. This is actually unsurprising (and I invite anyone to try this experiment on an unsuspecting audience). However, imagery such as this is the starting point of the larger meta-narrative that "technology" has a certain appearance, and is both the steering wheel as well as the engine that drives progress. This imagery compounds itself and, as Smith (1994b) notes, allures us towards into even constructing technology-driven explanations of historical processes, including cultural processes. What has also become naturalised is how technology has been placed as a crucial change agent in the culture of modernity, which has stemmed out of the ample visual and experiential evidence among people and communities from all walks of life who believe that technology, with the aura of an independent and autonomous entity, innately alters the ambience of everyday life (Marx and Smith, 1994). Despite abundant evidence that technology is not a sovereign entity and shares a reciprocal relationship to society, this naturalisation seems to remain (Bimber, 1990). Such an ascription of autonomic power to technology is *technological determinism*, a classic example of this being Karl Marx's conviction that the introduction of the railways in India would "dissolve" its caste system (Heilbroner, 1994). Other examples of this include placing the waterwheel as the founding artefact of manorialism, the steam engine as that of industrial society, or the stirrup that of feudalism (Hughes, 1994). The idea of technologies as driving forces of history was developed within the paradigm that regarded new technological artefacts as a sure-shot means of addressing social and politically defined goals catering to the more noble idea of social change (Marx, 1994, 1997, 2010). Smith (1994b), has pointed out that thought processes which planted faith and agency in technology as either a liberator or a annihilator can be traced back to the European Enlightenment, which gained further confidence in centuries of European colonialism that followed, and in the search for American political order. Heilbroner (1967) reveals that technological determinism is a peculiarity of a historical era characterised by high industrial capitalism. Driven by the temperament of the era, it emerged as a general conviction that cultural and social change, therefore, was steered by technology (Smith, 1994b), although there was scarcely a deep introspection on whether, by the same token, medieval technology brought about feudalism, or industrial technology capitalism, or electrification socialism (Heilbroner, 1967; Hughes, 1994).

An offspring of technological determinism is technological solutionism, which brings with it the mindset that technology holds the self-determined potential to ameliorate the human condition (Morozov, 2013). Technological solutionism employs a reductionist account of the analysis of development concerns to simplifiable "obstacles" that can be overcome, disregarding any sociological or historical complexity of either the concerns or the technological fix (Morozov, 2013: 14). This erroneous inclination is not exclusive to our era. Governors General of India Lord Dalhousie and Lord William Bentinck, back in the mid-1800s, believed that the railways and the telegraph could break down the caste system and usher in modernity into this wretched subcontinent by which the Indian lot could be civilised, saved, and pulled-up towards the industrialised standards of the West (Sarukkai, 2008). Arnold (2013) explains Harriet Martineau's vision which, much like Bentinck's, illustrated how the railways would expand the otherwise immutable nature of the Indian mind towards new horizons of thought and practice. Arnold also cites instances from Indian fiction, with an intriguing episode from Mulk Raj Anand's character – a manual scavenger traditionally obliged to handle human waste in India – viewing a flush toilet as an artefact that came with the promise of expunging caste from Indian society. Road transport and electrical technology were believed to be harbingers of social transformation and social harmony in the West, back in the nineteenth century (Morozov, 2013: 44). The idea, Morozov explains, is that the impact of technology (conceived to be essentially a neutral entity) naturally flows from its inherent qualities, regardless of context. Both technological determinism and technological solutionism overlook the social foundations of technological processes and outcomes, and severely compromise on multiplicity in conceptualisation, enquiry, and analysis.

There are a number of varieties of technological determinism, explained by Marx and Smith (1994). "Hard determinism," apparent in the label itself, ascribes an extreme scope of agency to technology, viewing advancing technology as an unstoppable force that determines the course of events. "Soft determinism," on the other hand, is the other equally unlikely extreme where sociocultural and political–economic forces hold all the agency in their firm grasp, simply denuding technology of all its power. Marx and Smith (1994) observe that popular notions and meta-narratives around technology incline towards the former, with people believing that technological advances "embody humanity's choice of its future" (p. xiv). Bimber (1990) presents to us more nuances of technological determinism, which include nomological (i.e., where the laws of nature drive an inevitable technological order, regardless of sociological or political agency) and normative (i.e., where technology influences history consequent to societies attaching cultural and political meaning to it). Yet another approach is labelled by Bimber (1990) as the "unintended consequences" account, which places primacy on the trait of uncertainty and uncontrollability of the outcomes of technology, despite the potency of social actors – in other words, technology is partially autonomous and its advances determine social outcomes beyond any human control.

An interesting standpoint is that technological determinism may not, in the first place, be about technology but may actually be a cultural and political fetishism where artefacts proxy technology and technology indicates national progress; technological determinism may hence be a cultural experience, framed by cultural preconceptions (Smith, 1994a).

The public indoctrination of technological determinism – rounding up some unusual suspects

Let me bring this to India, where this book is set. No single actor or body of thought is solely culpable for the oversight in appreciating the interaction between sociological and technological processes. The undercurrents of this disregard are practically everywhere in India – the private corporate sector, public discourses, technocrats, state media, and so on. However, to understand the malady in its entirety, it would be appropriate to begin at the top, glancing at India's science and technology policies crafted by the state.

At the top

India has had a spotty record of creating deep, visionary, and pragmatic policies around science and technology, despite the rhetoric around the role of technology in development that has only crescendoed through the decades. There have been a series of well-intentioned documents, often created by groups of highly accomplished individuals and esteemed organisations and institutions. These include the Technology Policy Statement 1983; the Science, Technology, and Innovation Policy 2003, and its supposedly improved version in 2013; besides sectoral policy documents that have sprinkled the role of science and technology across their pages. There have also been the National Digital Literacy Mission, the Digital India initiative, and the National Strategy for Artificial Intelligence. Over and above all these policy documents have been the Five-Year Plans over the last 70 years, many of which place technological change as a central actor in the development process in India. The intensity and flavour of the role of technology in development is uneven across the Plans, but a glimpse across them reveals a strong inclination towards technological determinism emerging through the years. Let us stroll through the Plans.

The First Plan (1951–1956) and Second Plan (1956–1961) actually made a good technological head start for a newborn nation, by advocating for the importance of research laboratories and building up national capabilities in science and technology, and consequently allowing for provisions for expansion and institutional linkages across universities and other organisations around the country. While this can be suspected to be the seed of technological determinism in the Indian development story, it actually may not be the case. For the economy and society that characterised the optimism of the 1950s, soon after Independence in 1947, it was probably essential for one actor to get the fire started – which was, for all practical purposes, the state. While prior to 1947, scientific research intuitions were already doing commendable work on furthering the cause of science, both for the country's sake and for science's own sake, a good-natured but firm rudder was probably necessary at the time to gently sway the ship to channel India's scientific abilities towards also playing a role in building technologies for its socioeconomic development. Consequently, the celebrated Indian Institutes of Technology were set up in Kharagpur, Bombay, Madras, Kanpur, and Delhi; capital goods and consumer goods industries employing high technology were encouraged; and in 1959, the Inventions Promotion Board was established to buoy up the "spirit of invention" among individuals ranging from artisans to technologists by providing technical and financial assistance. The Third Plan (1961–1966), however, is quite explicit, with a clear statement of how higher standards of socioeconomic life and the speedy pace of development in other, more advanced, countries are a *direct result* of progress in scientific and technological research, both pure and applied. This is reverberated in subsequent Plans, and begins to balloon in the Sixth Plan (1980–1985), which strongly emphasises that because science and technology are indubitably the "instruments of social and economic change," the development of these endeavours have to be placed as a major objective of planning. The Technology Policy Statement 1983 sets right at the outset within its preamble that science and technology are the very *bases* of economic progress, and possess the capacity "to solve problems." Taking a full numerical stock of the nation's laboratories, councils, universities, institutions, and manpower, the Sixth Plan makes a compelling argument that

> when we consider the magnitude and dimensions of India's problems of economic and social development, associated with the vast and increasing population and immense poverty, especially rural poverty, it becomes clear that massive application of science and technology has to be an essential component for their solution.
>
> (Sixth Five-Year Plan, section 19.6)

However, due credit must be given to this Plan (as well as the Statement) for acknowledging fully that these ambitious tasks must be accompanied by an economy and society that places a premium on the inculcation of scientific temper in its citizenry, and that sociocultural factors are to be assessed side by side with technological considerations:

> the task of creating a scientific temper is a vital necessity for the growth of science and its utilisation in the development process. There is need to create a scientific climate and involve the people in discussions on various issues of science and technology which affect their lives . . . there is also need for promoting public debate on major issues of science and technology. The full potential of science has to be utilised for eradication of irrational attitudes, which tend to hold back the nation from the path of progress.
>
> (Sixth Five-Year Plan, section 19.9)

Special sections are accorded in this Plan to involve disempowered social groups, such as those communities listed under the Scheduled Castes and Scheduled Tribes, as well as to highlight the historically neglected role of women in the science and technology process (though not without sexualising occupations, apparent in a list of fields of technology that would be "appropriate" and "beneficial" for women). Even until the Sixth Plan – which was modelled within the decade that saw a gradual economic and industrial liberalisation process in the economy – despite a ballooning of the role of technology in development, there was some feeble yet audible appreciation of the sociological connections this role inevitably makes in the Indian context. However, in an overall assessment, it is clear that this appreciation rarely went beyond a mere mention.

From the Seventh Plan (1985–1990) on, the Plan narratives steadily deepen the idea that it is technology that is in the driver's seat, while the rest of economy and society are its obliged beneficiaries. The roles that the latter play, if any, are to create appropriate conditions for technological self-reliance and to better employ existing technologies, suggesting that technology and socioeconomy may share tight relations but are still two separate entities where the technology sphere may by *assisted* but never modelled by the socioeconomy sphere, for the latter's propulsion ahead. The Eighth (1992–1997) and Ninth (1997–2002) Plans amplify the momentum significantly, by calling for more innovative activity and for greater permeation of technology (read, *digital* technology or *sophisticated high* technology) into spheres such as transparent governance, citizenship, rights, rural development, disaster management, and so on. By the time we arrive at the Tenth, Eleventh, and Twelfth Plans, the information technology enabled services (ITES) sector appears to take centre stage almost as a universal panacea to negotiate solutions within and around the previously named spheres, so that in the process India may "develop" further and march towards becoming an "IT superpower." There is, as we move ahead in the Plans, a diminishing regard for grounded processes (apart from token and ornamental mentions of grassroots innovations with a highly paternalistic viewpoint), and a ritual repetition of the term "local" in the technology-development relationship cast in a tone evoking passive reception instead of active agency. Overall, the spotlight intensely illuminates on science and technology as deliverers of development, while the spotlight on their social context of functioning gradually dims out.

The Science, Technology and Innovation (STI) Policies 2003 and 2013 are not too different in their inclination, citing "context specificity" and "local factors" purely as incidental mentions, besides the occasional "inclusive growth" platitude. While terms such as "inculcating scientific temper" appear every so often, their treatment in the policy documents is shallow at best, or they are interpreted as quantitative increases in science and technology manpower. The role of women in technological outcomes is entirely ignorant of the complex historical relationship between gender and technology, and pruned into simply providing incentives to young, talented female graduates towards a science and technology profession. Overall, the push seems to be for accelerating scientific discoveries and technological improvements in the quest for "faster, sustainable, and inclusive growth" (STI Policy, 2013: 17). The National Digital Literacy Mission and Digital India also display noble pursuits of making one person in every family digitally literate, and to impart information technology training to them (i.e., to improve their computer operating skills), "so as to enable them to actively and efficiently participate in the democratic and developmental process and also enhance their livelihood" (NIELIT, 2019). Finally, the National Strategy for Artificial Intelligence 2018 (NITI Aayog, 2018) thrusts forward the proposition that while artificial intelligence may be viewed as a disruptive force or unstoppable tide, it can certainly be transformed to provide assistance in leaping strides in the social sector, such as in the spheres of education and healthcare. Once again, society is placed as the passive recipient, and not a forger of technological change.

Every single document in the august list earlier has unmistakably positive intentions in channelling technological change for the greater good. All of them tend to match their wavelength with the global discourse on the role of technology in development, and wish for greater technological participation in shaping India's development trajectory. Anyone might wonder what I was witch hunting for while critically reading through these policies and programmes – it certainly was not to investigate into the technological feasibility or the availability of supportive economic infrastructure to cradle these goals, and neither was it to craftily look for something outside of their epistemic ambit. My argument here is that their honourable intentions may never see the light of day in the broader inclusive sense of development because, even if these programmes or Plans are not apparently untied from immediate or long-term concerns of development, they are certainly short of crafting their ambitions within the gigantic jumble of socioeconomic realities that India is. Does this mean they should have had diluted goals or should have positioned themselves pessimistically or coyly? Probably not. What they ought to have done is reassess their goals, or at least recalibrate them, bearing in mind that the development and subsequent operationalisation of technological ambitions for facilitating development are not context-independent, linear streamlined processes. All goals and designs around technology and development are necessarily grounded in outcome and are heavily context-specific in deployment. Does that mean that programmes work only at a micro or meso level, and that there cannot possibly be a national policy on technology and development? I propose that there can indeed (and must) be such a series of macro policies, but the deployment of their content ought to be modelled based on the sociological terrain of each site, and that this fact ought to be conceded by the programmes and Plans. I do not allege that these programmes are in denial of obstacles in their path. I do put forward the accusation, however, that the obstacles confessed within the respective documents are mostly technical or mainstream-economic, and completely fail to observe that it is entirely plausible (almost certain) for the technological mission and its conjectured results to stand neck-deep in the dense sociological realities in India within each context.

Motion pictures and reflecting attitudes

The indifference towards the society–technology nexus from the part of the state can also be witnessed through a series of interesting short films that it created from the 1950s to the 1970s, including the aforementioned *Symphony of Progress*. It might seem awkward, probably even irrelevant, that I choose state-crafted short films as one of my bases for assessing the penetration of technological determinism and solutionism into public perception. However, I argue that these sources are far from simply ornamental entertainment pieces, and are deeply reflective of an attitude around the role of technology in development – paradigms that characterised spheres ranging from policy stance, to popular discourse, to cultural diffusion around perceptions of technological experience and outcomes. The state may exhibit its stance even in cultural artefacts such as films. These films are freely available online due to the proactive efforts of the Films Division of India, a part of the Ministry of Information and Broadcasting of the Government of India, which has in its long history produced nearly 10,000 such short pieces on a spectrum of themes. In an era long before even single-channel state-controlled television, which was itself a powerful medium of state propaganda, these films served as in-house documentaries showcasing the state of economy and society, and the central role of technological progress in their trajectories. Probably influenced by Soviet antics of state propaganda through motion pictures, these films showcase vignettes of some of the thrilling achievements of socioeconomic progress in India soon after the Plan era took off in the 1950s, allegedly through thrusts in big technology (much to the chagrin of those who sided with Gandhi's contrasting scale of operation); some even went as far as to provide evidence of a certain tenor of scientific temper permeating through the economy. A visit through these films is certainly worthwhile.

The very titles are eye-catching – *A Village Smiles, Build Machines Build India, Saga of Progress, Our Industrial Age, Electricity in the Service of Man, Destination Technology,* and so on. The background score of Indian music interspersed with Wagnerian overtones aptly resonates the optimism supporting the strong assertive narration of a male voice. The pace of the films is calm and moderate, but they unabashedly portray a sleeping giant awakening. They exalt the idea of an India in transformation, from an archaic and wretched rustic existence to a richer and progressive realm, achieved successfully by the overwhelming tide of big technology, which tames nature and propels society simultaneously.[1] The task, they say, is in the hands of young people – "students of progress" – driving the engine by engaging in subjects such as planning and technology (*Symphony of Progress*), which are domains of study that are "unfettered by tradition and in tune with the modern age" (*The Five-Year Plan in the Eastern Region*), supposedly the surest path for turning economic aspirations into reality (*Industrial India*). However, "technology" is usually large-scale infrastructure and industry, connected to dams or mammoth capital goods industries. Only big technology builds India, as it facilitates happiness among peoples who hitherto suffered the vagaries of nature, and helps humankind gain control over its fate (*Hirakud; Build Machines Build India*). It is truly a thing of beauty how big dam technology facilitates other big technology (*The Five-Year Plan in the Eastern Region*), and in due course, with a 100-percent guarantee, brings a smile to a village, sows new industry, creates new employment, and spawns wealth for millions (*Hirakud, Symbols of Progress,* and *A Village Smiles*). From their part, the recipients of these treasures must discard obsolescence to support the inevitable wonders of big technology and work and life crafted around it; illustrated in film with juxtaposed clips of depressed or barren land versus dams and heavy industry, mosques or temples versus technological marvels (*A Village Smiles*). Surely, research institutions emerge as the outposts of the new economy, and hence, scientists and technologists arrive as the alchemists of the modern age (*The Five-Year Plan in the Eastern Region*). Why only industry and agriculture, the detrimental effects of even demographic epidemics such as overpopulation can be overcome by one combination – technology and speed – repeated thrice in the film *A Village Smiles*: "speed and technology are the only answers." "Electric power [read, big technology] is the 'open sesame' to a more pleasant and prosperous life" (*Electricity in the Service of Man*), where all it requires is a "flick of a switch." Modernity, progress, prosperity, well-being, transformation, agency, development – a medley of gilded terms pockmarks all these films – powered by the fuel of technology and a big-technology–oriented economy. The new temples of modern India "fight to dispel the darkness" (*A Village Smiles*), which stand in as thickset brawny and highbrow testimony as the exalted Sun Temple at Konark – therefore, Konarks of our own era (*Hirakud*).

Large capital goods industries were ineluctable at the time, given our economic and infrastructural situation. Dams, while indeed engineering marvels, have notoriously released torrents of social and ecological strife and searingly painful displacement from their sluices, as much as they did water. My point here, however, is to observe how they – and by proxy, technological change – were assumed to immediately and *automatically* usher in modernity that would ripple though society and economy in creative destruction. Technology was deemed the indisputable juggernaut of our contemporary era, advancing in majesty with bureaucrat and technocrat as propeller and rudder, flattening anything archaic or unworthy in its path. Entirely unaccounted for are any of the social variables that might interfere with or shape technological progress, to bring about the prosperity reflected in these films. I consider this oversight a mirror image of much larger attitudinal paradigms around technological change and development.

Of cyber-coolies and missile-men

Moving ahead to the 1990s and beyond, these attitudinal paradigms also pervaded in the larger discourse around an emerging sector that captured the imagination of an entire generation of young urban English-educated individuals, their families, and their peers – the information technology enabled services (ITES) sector. The emergence of this sector in India provided tremendous hope to young engineering graduates and other urban English-educated undergraduates that they could not only be part of an industry that became synonymous with the liberalisation–privatisation–globalisation experience in India but also participate in building new ideas of modernity, vastly apart from big dams or public sector industries. All through the 1990s and the early 2000s, the ITES sector in India boomed and became the supreme icon of a hopeful young and new India, a freshness that was missing for some decades in the country. In contrast to the starry-eyed hope of a newly independent India that generations prior experienced, this was a vision of a modern technological era, something much bigger than just economic self-sufficiency, because it also displayed the potential, perhaps even with a guarantee, that India would flourish and find its place of prominence in the world economic map as a beacon of technological power and progress.

Technology this time was not a behemoth Konark temple or a stomping and trudging heavy industry. Technology now was capsulated in a sector viewed as sleek, streamlined, and providing a liberating ambience in life and work that allowed young people to dart about in their career trajectories, realising dreams for themselves and the country. The sector made a covenant with the new generation that it was no longer corrupt and nepotistic politicians or self-serving bureaucrats who would shape India's destiny, but the *youth* who would now define modernity and a progressive India where family legacy or political influence was passé, and "everyone" could come on board in the new technological era. A brilliant anthropological study of the sector by Carol Upadhya in 2016 describes in fine-grained detail how "software engineering" as a career choice in India spawned around celebratory narratives about the ITES industry as brimming with modernity, cosmopolitanism, and upward economic mobility. Coverage of the ITES sector in the popular press and television media carried a whiff of pride that this was India's "moment." Popular culture radiated the sanguineness that the sector promised in terms of an enduringly upwardly mobile life. Of course, "software engineering" was a colloquial and loose catch-all term for any professionally trained association with computer systems (an enormous range of economic and scientific/technological activities), which reflected the uninformed popular discourse around the actual nature and heterogeneity of work within the ITES industry. Upadhya (2016) illustrates how all these framings were embedded within tropes around the "middle class" – a category that remains vague but is proudly exhibited by individuals as a badge with a veneer of high morality and upward mobility. She continues explaining how the role ascribed to the ITES industry was unabashedly linked to the popular idea that by means of new technologies, India would "leapfrog" over the usual stages of economic growth to become a "knowledge economy" (p. 38). Masterfully, Upadhya also argues how the formulation of a unique digital citizen identification system, such as the Aadhar overseen by high-standing individuals from this industry (and not other sectors), was not incidental, as it represented the seepage of solutionism and technological fix from digital-technology–based industry to public administration and "modern" governance. I extend her argument that this was actually a symptom of the attitude that anything and everything in grand scheme of things needed a digital or high-tech solution that emerged from the new site of India's modernity and progress – the ITES industry. Not only in public policy and administration but even in the everyday life of common people across urban and small-town India, this industry and the aura of *technology* it imparted had profound effects. Getting a job in this industry was considered the ultimate success, as it not only meant a high-salaried occupation but also that one would be a part (even if a cogwheel) of the magic potion that would make India sparkle. A career in ITES meant a technology-embedded life, too. An "IT-job" was a metaphor for a certain genre of life and work. Hence, Upadhya notes that the industry and its heavily laden metaphors of technological determinism/solutionism bound together two trajectories that are often construed as opposed to one another: individual progress and national progress. The twain finally met, and there was little looking back, because it was not just that technology would take India towards its future . . . it *was* our future.

Eventually, and naturally, word began passing around social circles both casually and alarmingly that the work-related physical and mental stress and detrimental spill overs in life and living associated with this profession, had never been witnessed before in urban Indian economy and life. Academic study and careers in engineering and technology began to steadily be perceived as a crowded boxing ring (with no decline in enrolment into the ring), with stories about erstwhile software engineers shifting to more creative and fulfilling pursuits in life (such as photography or theatre) abounding in the popular press and popular culture. More importantly, the reality of the sophistication of work in the Indian ITES industry when placed in a global context was very low-end, and the reality was glaringly out of tune with the dream that this was how modern technological India would look. Great conviction around an ITES-led economy turned sour, and the sector slowly emerged as synonymous with long and pointless work and a professional rat-race unmatched in independent India's economic history. Public discourse around the sector increasingly began to display sentiments of precarity, which was not surprising, as the ITES industry stands proud as the epitome of a neoliberal work culture.

It is worthwhile to reproduce from Upadhya and Vasavi (2006: 158–159) for our general understanding of the hollowness of the "information society" or "knowledge economy" portrayal based on the ITES sector. Especially in the popular imagination, a city such as Bangalore has long been considered a beacon of the ITES industry, with nomenclature such as "Silicon Valley" routinely attached to it. Employment opportunities in this industry built on existing class divisions rather than absorbing a broader cross-section of the population. The industry appeared to have drawn primarily individuals who have attended elite educational institutions, and who hail from the educated urban middle classes. The vast majority of the subaltern, urban underclass, and rural people were not absorbed by the ITES industry – which is not of course deliberate, but due to the fact that the requisite skill set and cultural and social capital is generally out of reach for the greater populace. While the industry as a whole may have procreated ancillary service industries, the polarised socioeconomic difference between IT workers and local population in peri-urban Bangalore was apparent, and has generated social tensions. Dualism is only too evident in these regions, if one ever visits them.

A slow but steady disillusionment with this industry grew, but the mindset around technology, in principle, did not budge. That is, the notion around ITES as "India's future" had begun to fade, but the notion that technology would still drive progress continued. The damage had already been done, and the idea that technology-based sectors (even if not ITES) and high technology would be the future of socioeconomic development had already seeped into the urban Indian mind, deepening the technological determinism and solutionism mindset.

In the era when the ITES was celebrated as *the* industry that would shape India's technological future, an untypical development materialised on the Indian political scene that brought cheer even to those among the public who claimed to generally stay away from politics. For the very first time, an august office that was usually graced by scholars and lawyers was now occupied by a technologist who won the position almost unopposed. APJ Abdul Kalam assumed office as the President of India in great pomp and circumstance in 2002, and his affiliation to the sphere of missile technology was viewed as a shining credential during an era where India was aspiring to become a "knowledge superpower." Not only was Kalam immensely popular as a president who connected with common people with (often amusing) flair, he also penned a series of books and other writings that set down the agenda for a "developed India." While due credit must be given to the late president for presenting a vision of an "aspiring India" at a time when there was almost no one else who delivered a performance as India's dreamer with the showmanship that he did, I take this opportunity to pose a strong critique not of his body of written work in general but of his perception of the role of technology in his agenda for India's development. With this, I am fully aware that I run the risk of playing spoilsport with some of the most popular inspirational writing on India's developmental aspirations, which genuinely kindled the minds of millions of adults and children alike to envision a better India.

I refer to three of his popular works, *Envisioning an Empowered Nation: Technology for Social Transformation*, *Ignited Minds*, and *India 2020* (i.e., Kalam and Pillai, 2004; Kalam, 2002; Kalam and Rajan, 2002), not as his best works but as those that represent his popular writing on development. Clearly, technology is placed as the "prime mover" for the "developed" nation that India ought to surge to by the year 2020. Technology is the entity that "dominates all walks of life." Kalam and Pillai (2004) attempt to chart the complex link between technology and society, by proposing that science is linked to technology through applications (technological, I assume), which is in turn linked to economy and environment through industry, by which technology and society are finally interlinked. A second Green Revolution must ideally feature in the agenda in the technology–agriculture interface, but most of all, what is required is a transformation of the existing society to a "knowledge society" by spawning more "knowledge workers" who can create, nurture, and exploit new knowledge for two purposes – wealth generation and societal transformation. Put bluntly, Kalam's writings sketched a unidirectional linear causation from technological change to social change, without either considering the reverse, or its harmony. A cursory look at the thrust sectors for the purposes of social change and wealth generation divulges the mind of the technological determinist and solutionist further – information and communication technologies (ICTs), biotechnology, weather forecasting, disaster management, telemedicine, tele-education, native knowledge production, service sector, and infotainment, all linked by information technology. To make matters worse, the writing calls for the reader to reflect on the various technologies that may "come in a big way," or "become obsolete or disappear," and most of all, which can "dramatically change Indian social or economic conditions" (Kalam and Rajan, 2002: 53) – as though any of these three processes are predictable, more so by laypeople. Technology is supposedly "the highest wealth generator in the shortest possible period" (Kalam and Rajan, 2002: 48), which strengthens all other "structures" of the nation.

At best, this writing was uninformed. At worst, it misled millions of adults and children. It had no place for the role of society, except as a ductile recipient of technological change.

Thankfully, most of these writings were restricted to the popular sphere and only marginally influenced technology policy. Unfortunately, though, they had permeated rather deeply into the public imagination, convincing people that technology was not just a juggernaut (a reflection of the thinking until the 1980s) but now a *silver bullet* that would address development concerns and *solve* them. Everyone from school children to young adults to working professionals had glowing admiration for Kalam in his outreach that is now the stuff of legend, and had sufficiently absorbed enough of his writing to thicken the determinism and solutionism broth. And at a subliminal level, technology emerged as an actor that improved individual, societal, and national well-being; not just in the driver's seat holding the steering wheel but powering the engine and the wheels of the car as well.

Kalam's writings and speeches on the theme are never solely responsible for technological determinism and solutionism to penetrate into the Indian consciousness, but I have mentioned them here because they played a compelling role in manufacturing a "technological era" in the public consciousness both in the present and the future. The implications of this must be meditated upon.

Expectations and disappointments

The public imagination around this manufactured "technological" or "digital" era was witnessed in several spheres. Anything that began with the prefix "e-" (implying "electronic" or "smart") immediately began to rake up expectations in the minds of the public to "solve" with a point-and-shoot strategy. Most issues around development began to be aimed at by digital technological interventions – in education, environment, polity, infrastructure, health, communication, and so on – with results that wavered between successful, indifferent, and dangerous. Some, such as linking the Aadhar programme with public distribution of essentials such as food rations, had disastrous consequences with even children of marginalised communities dying of starvation (such as in Jharkhand state across 2017 and 2018) because their families could not "identify" themselves digitally and claim their food entitlements. In principle, there is little wrong with a technological intervention to address a development concern. The problem was, and remains, force-feeding of technological intervention (usually digital) to socioeconomic processes as not just a high-protein supplement but as a *panacea*. The goal of diffusion often becomes the objective superseding actual development concerns such as public food distribution or education. The aura of technology as a solve-all pervaded even the development discourse in policy and state planning with the normalisation of "technology-led" development, as opposed to deploying technological interventions as options among several other possibilities, subsequent to a nuanced understanding of the role of technology in development according to context.

Expectations and disappointments bumped into one another along the path that was promised to be sleek and towards a shining destination in the horizon. ICTs appeared everywhere, in the form of mobile phones that were bought and operated by individuals in all walks of life and in *all* strata of economy and society; but while some benefited from them in livelihood and opportunity, using them as a springboard, for others they were nothing more than an entertainment tool or for casual conversation. Digital technologies were everywhere but absorbed unevenly by different social groups and across economic strata. "Smart" things appeared everywhere but not for everyone, often excluding people and communities systematically when encountered.

At some level, we might have as a populace been confused about why "technology" was not exactly taking us places, and why there was a coexistence of expectation and disappointment in the development experience despite technology being "everywhere." The reason for all this disorientation was not simply because the technology was inappropriate or insufficiently developed, or that diffusion of effective technologies was stunted. It is certainly not the case also that we have inadequate manpower, institutions, or resources in India. The reason for the perplexity is quite simple: the idea that we were (and are) living in some sort of "technological era" since the 1990s, based on plain-to-see evidence that digital technologies were suffusing through nearly all arenas in development and everyday life, was a manufactured epoch. It boosted unnecessary optimism around technology, without realising that technology seldom operates in isolation and is crafted by the society it resonates in. This "era" was created by a reckless concoction of the factors and events described in this section, and more.

Where were the academics?

A space that has questioned all this, even if not directly influencing popular and policy discourse, has been the scholarly literature in India. Academic writing on the role of technology and development in India was mixed in its perspective. While there is an enormous body of literature in India around a conceptual framework such as national systems of innovation, and the interlinkages between technology and Indian development are as old as Indian academic writing, the Indian literature on this subject has demonstrated an awareness that neither is technology going to "drive" everything or simply "solve" everything, nor is it devoid of social context. That is, the idea that society shapes technological experience (as much as technological change repositions certain facets of society) was appreciated. While the Indian literature has not launched a dazzling attack on technological determinism or solutionism, it has nevertheless embraced the fact that there is a social context to technological experience or outcome. While it would be a gargantuan task to provide a review of the entire literature on technology and development in India (and which may perhaps be needlessly vast for this book), I have undertaken a small exercise of surveying through 50 years of a publication that is possibly the flag-bearer of Indian academic writing on issues of socioeconomic development – *Economic and Political Weekly* (*EPW*). While restricting to one journal is undoubtedly inappropriate while surveying literature on a theme, I take the liberty of venturing into a task that may be methodologically incorrect but metaphorically appropriate. In other words, while the Indian literature on technology and development spans across dozens of journals and book volumes, I choose *EPW* because it greatly (even if not entirely) reflects the tenor of Indian academic writing on socioeconomic development.

I surveyed articles in *EPW* from 1967 to 2018 that deal directly or indirectly with technology and development. They range from issues of export and import of technology, to capability building, to industrial and agricultural technology, to infrastructure and institutions, to rural economy, and of course policy. It would be mechanical to provide a summary list of articles that nod[2] at the idea of a social context of technological experience, or even to provide a list of articles that overlook it. The manner in which this body of thought and evidence demonstrates its more inclusive stance towards "social context" is not in the usage of this exact term; neither is the literature here a critique of technological determinism in the manner displayed by a great volume of writing from the West. The literature has, instead, suggested with compelling rigour that there are ideological and social concerns in the progress of technology in India (Sinha, 1967), and that questions of technological choice, for tackling underdevelopment, are indistinguishable from broader questions of social transformation (Vyasulu, 1976). Sociological concerns are not only at ground level but even at institutional level, such as concerning the social makeup of the Indian Institutes of Technology (Rajagopalan and Singh, 1968). It is lamented that the sociocultural context of science and technology in India is yet to be explored and analysed, as it determines how technology is ultimately operationalised and deployed (Ahmad, 1983). Ahmad, in his coverage of a national seminar on the social relations of scientific and technological change, goes to the extent of even asserting that the impact of society on science is in "far greater measure" (p. 1561) than the obverse. On the basis of the premise that social concerns significantly shape technological concerns, Bhalla (1987) asserts that technological progress is far greater than simply a "technological" issue and is an institutional, social, and political one, hence technological change has to be examined in tandem with social and economic change, while calibrating development strategy. Science and technology policy has simply omitted these considerations, including those of gender and women's concerns, which are sometimes glaringly seen even in the vocabulary used, let alone the slant of patriarchy in technology policy (Swaminathan, 1991). In contrast, people's movements in science and technology have appreciated and embraced the realities of its social character (Mohan, 1995; Varma, 2001).

Towards a social context of technological experience

With this background on a technological determinist and solutionist mindset in India, it is imperative to now move towards a conceptual discussion of the fundamental idea of society–technology overlap. The first step towards embracing such an understanding is to reiterate what I've been discussing from the outset of the book – that the stuff of technology is much more than science and engineering, and is sociology and politics as well (Bijker et al., 1987; Bijker, 1997, 2010; Bijker and Pinch, 1987). Let me first state that I do not wish to drive the basis of this book towards social constructionism or cultural determinism in its strict sense.[3] While the former puts society as supreme, the latter implies that technology is nothing but a functionality of human society (Sarukkai, 2008). However, given the pervasive tendency of technological determinism and solutionism in influencing economy and society in contemporary India (as seen in the previous section), it is much more important to tackle this malady than its sociological extreme.

To excise the notion of neutrality or indifference of the entity of technology, we begin from the seminal work of Langdon Winner (1980), who proposes that the creation and operationalisation of technologies are a factor of the societies they are embedded in. MacKenzie and Wajcman (1999: 5) capture the essence of his work, in that (a) technologies are designed to expand certain social options and restrict others, and that (b) technologies are political, not only in their design but actually in their entirety; this therefore implies that technologies are more compatible with some social relations than others. Continuing on their own exposition, MacKenzie and Wajcman (1999) explain how by adopting a perspective of technological neutrality, we end up divesting from public discussion the very nature of technology itself as a feature of political life; after all, they state, by moving towards an understanding that technology is a product of the market, we seem to convince ourselves that the market is the ultimate social institution. Sarukkai (2008) adds, similarly, that the characteristic of non-neutrality in technology is what brings differential results across cultures, which implies that we must depart from the stance that technology is functionality and nothing else. The move away from this, formulated early on in the literature by Winner (1980), involves rectifying the tone, tendency, and content of the public discussion around technology from one that studies unidirectionally the "social impact" of a technology to one that studies the social *circumstances* of its development, deployment, and use. The latter calls for far greater interdisciplinarity, as well as conceptual and methodological multiplicity in understanding and analysis around technology. It also brings out critical undertones of power and other indoctrinated social institutions and structures that may be tightly linked to certain technologies (Winner, 1980).

One of the theoretical conceptualisations on this is *technological momentum* by Thomas Hughes (1969, 1994). This is an alternative that builds on the shortcomings of both technological determinism (of Karl Marx, Lynn White, and Jacques Ellul) and social construction (of Wiebe Bijker, Trevor Pinch, and interestingly,

18

Hughes himself), where both forces are in operation but swayed by time. That is, while social determinism may impose its influence on a technology in its youth, technological determinism tugs at it as the technology matures. Hughes therefore allows for enough complexity and flexibility of deploying either society or technology as the drivers in historical processes, depending on time and context. This theory implores us to adapt it, especially in the understanding of big technologies – maybe even to our temples of modern India?

Hence, what we need here in our understanding is to depose technology from a *primum mobile* to a mediator that overlaps with society, which doesn't necessarily reduce the actual impact of technology but on the contrary, improves our fine-grained understanding of its operation and outcome (Heilbroner, 1967). If we wish to build a more holistic understanding of technology, we are pressed to integrate and highlight social and cultural factors – such as gender, power, and race – into our analyses of technological development and experience, and thus to improve our understanding of the human condition and its relationship with technological change at a broader level (Lerman et al., 2003). What I imply here, borrowing from Scranton (1994, 1995) (who in turn, adapts the work of Mark Granovetter), is that both are embedded in one another and one is not a "dependent variable" within the other "autonomous process." Society and technology form an intricate socio-technical network, with the whole at the macro level typified by the complexity of its constituents at the micro level (Misa, 1994). David Arnold's brilliant exposition (Arnold, 2013) brings this to life, by masterfully demonstrating the society–technology intermesh at a micro/everyday pitch, without losing sight of the macro picture in terms of the outlook towards technology by independent India's early policymakers. Picking up what he terms "everyday technologies" such as the bicycle, sewing machine, typewriter, and rice-mill, Arnold showed how Indian society was, first, not as rigid towards technological change as Martineau cast it to be, and second, that both the diffusion of the technology in the Indian landscape and the societal and political trajectories around these technologies grew together. He also provided evidence on how technologies that formed the everyday life of the West as much as in India had dramatically differing experiences in the Indian context, rooted in the ingenuity of the social interaction they had here; in essence, their social ownership. Arnold thus calls for the "social life" of technologies to be understood in a much deeper sense than simply the diffusion of those technologies. This also is in contrast to the oversimplified success/failure dichotomy of technological experience, which is too often reflected in policy.

There is no singular theory of the social context of technological experience. In fact, having such a singular theory would be self-defeating. Each technological experience has its own peculiar assortment of concepts and approaches, and local actualities, in appreciating its social context. In this book, I demonstrate exactly this. I move ahead with Arnold's clarion call for understanding how technologies, though they might have materialised as identical artefacts across the globe, still have "local uses and vernacular meanings" (Arnold, 2013: 4) that might supersede these global templates. I proceed with the reality that technological experience is not only a metropolitan phenomenon but also has subaltern and gendered elements, in ambiences wildly different from corporate white-collar or sophisticated industrial settings, and which can be fathomed only with a multiplicity of concept and method.

The task ahead in this book

When I present selected technological experiences in contemporary India, my objective is neither to establish any new theoretical propositions nor to hinge on any established ones. It is neither to unquestioningly bank upon the virtues of a largely qualitative methodology, nor to depose the rigour of the quantitative or statistical methodology. My goal here is to propound the importance of *multiplicity* in concept, method, and subject when dealing with socio-technological experiences, resonating with the nature of their variability. To drive home the point about multiplicity in local contexts and vernacular meanings, I wish not to lay out a string of anecdotal instances that provide evidence for social context in technological outcomes. Here, I provide three case studies. Each of these cases are vastly different from one another in possibly every aspect, except to substantiate the society–technology nexus. With these case studies, I also promote a perspective for policy and public discourse around the role of technology in development, as well as the reality of technological experience. I present the three cases in this book in chronological order of their empirical study:

1   On caste weavers in a small town;
2   On deprived communities in a peri-urban setting; and
3   On women garbage-collection informal-workers in a metropolitan city.

The first case study in this book is the case of the Saliyar caste-weaver cluster in the town of Balaramapuram, near Trivandrum, Kerala, producing handloom textile products. This study demonstrates an interesting form of technological change and learning, i.e., through informal information sharing among interpersonal social networks (these networks often defined by caste in India). Lessons drawn from the Saliyar cluster's experience show how thickly cohesive social and production networks (resting on the Saliyars' historically rooted social and ethnic capital) have broken the long-standing dominance of this community, both economically and technologically. The methodology employed in these two chapters is network analysis, both quantitative and historical-qualitative. By undertaking a network analysis of the Saliyar cluster, I provide evidence that it is not just embeddedness alone but in its *combination* with homophily (the preference for one's own kind) in various intensities that is detrimental. The study demonstrates how the conceptual ambit of embeddedness has to broaden to recognise that social relations come in various "homophilies" (with a novel term – "homophilous embeddedness," the conceptual contribution of this study);

and in turn, what consequences this cohesion has for technological and economic outcomes. I continue to demonstrate that while there are a multitude of cases in Indian handloom history demonstrating a healthy relationship between community cohesion and technological progress among handloom weaver communities, in the case of the Saliyars the relationship has been antagonistic and unhealthy. To understand why, I investigate into the inherited nature of Saliyar networks, the centrality of community social capital among the Saliyars and, most importantly, the inherited cohesion in their networks. This study also presents how complex social relations influence economic relations and technological progress when these relations are relayed across generations.

The second study in this book presents a fresh perspective on the complicated relationship between digital communication technologies and historically disadvantaged castes such as Dalits, in peri-urban Bangalore. The study deepens the understanding of the experience and impact that mobile phones have had on the lives and livelihood of these historically deprived castes. The research is set in peri-urban Bangalore city around Electronics City, pictured by policy and popular perception as India's "Silicon Valley." Primary oral accounts from interviews across Dalit households, Dalit entrepreneurs, and Dalit political activists are read against conceptual and empirical perspectives in the literature on digital divide and information society. With these accounts, the study asks whether mobile phone technologies have simply bypassed or been insufficiently harnessed by Dalits in the region to overcome historic deprivation, and also whether they have even assisted in the reinforcement of exclusions for some of these groups. I use a modified form of the oral histories method, as well as focus group discussions and detailed interviews, to answer these questions and thereby demonstrate how the contemporary socio-technological outcome among Dalits in peri-urban south Bangalore is a result of a convergence of three elements – (a) the durability of caste in peri-urban and metropolitan India, (b) the social construction of the usage of mobile phones, and (c) the myopia in the conventional understanding of the digital divide in India by policy and public discourse. As the literature has indicated, if society reproduces its cleavages in the various kinds of technologies that it uses, we have to adopt a new perspective of understanding the relationship between new technologies and the subaltern. We have to rethink the digital divide as not simply a binary of haves and have-nots but as a multi-dimensional spectrum, much like caste itself is a spectrum of inequality and forced deprivation. This chapter urges for a reassessment of our solutionist perception towards the revolutionary potential of mobile phones, an approach that has manifested in state programmes such as Digital India and the National Digital Literacy Mission.

The third, and final, study is about women garbage collection workers, or pourakarmikas, in metropolitan Bangalore city, India. Undertaken along with two co-researchers, we assert that gender stands as the cornerstone of their everyday work experiences as well as their everyday technological experiences. That is, the politics of informality and technological experience are rooted in notions of gender and labour informality. For Bangalore's garbage collection workers, a subaltern informal feminised workforce in an urban metropolitan setting, while work experiences are outcomes of a dismal concoction of multiple social variables (such as informality, caste, and gender), their encounter with a new technological intervention has reflected and reproduced gendered informal-labour experiences, and has simply seated itself in the existing sociopolitical ambience. Our analysis brings together conceptual traditions in labour informality among women workers and the feminist politics of technology, placing gender as the centrepiece to intertwine them, hence providing a cooperative analysis of informality and socio-technology. In going through this study, it ought to be no surprise how everyday work and technological experiences are outcomes of an amalgam of multiple social variables that have conjointly built the deeply downgraded footing for these female informal workers.

The final summarising chapter reexamines the motivation of this book, and provides a clarion call for furthering the research and policy agenda on the economic sociology of technological change and technological processes.

As I had admitted at the opening of this chapter, all three case studies appear as morose tales. However, in the final analysis, I aim not to provide successful models but to disentangle the technology–society nexus in an Indian setting, and present the assortment of theoretical concepts and methodological tools that I have employed to reveal the depth and intricacy of this nexus. In addition, I contribute further evidence on how innovations such as industrial technologies, communication technologies, and identification technologies are, as the literature has also thrust, not only about strides in science and engineering but also about politics and sociology. I demonstrate how an interplay of an array of sociopolitical factors can generate unexpected or often imperceptible technological outcomes, thus dethroning overconfident technological prescriptions that solutionists have built with their sweeping hagiographic view of technological change in general. I wish to present a much broader perspective on technological experiences and processes, as embedded in sociological underpinnings, compared to the stunted perspective of the general mainstream approach towards science and technology and innovation for development in India. I project innovation and learning as more inclusive processes, a perspective essential while crafting development policy. By subscribing to such a holistic and multidisciplinary approach, maintaining social structure and social relations as central to technological change, I would like to compel readers to paddle across a spacious span of perspective, method, and concept, to strengthen our understanding of the relationship between technology and society.

Technology has always been an integral part of being human. Archaeological evidence of sewing needles and agricultural practices tens of thousands of years old, experiments in metallurgy over thousands of years, the journey of print and information dissemination over the last few hundred years, and developments around energy over the last many decades – have all characterised each age in human history. We have *always* been in a technological era as humans, and there is nothing special about our time right now, except that the clamour of the messianic role of technology as an autonomous chauffeur of development and the first go-to magical solution has been getting raucous.

Notes

1 Most phrases in this section have been cited directly from the films to demonstrate their vocabulary and tenor of narration.
2 A sample of these is as follows: Rajagopalan and Singh (1968), Parthasarathi (1969), Vyasulu (1976), Ahmad (1983), Bhalla (1987), Swaminathan (1991), Wahyana (1994), Kannan (1994), Rath (1994), Mohan (1995), Banerjee and Mitter (1998), DN (2001), Varma (2001), Prasad (2005), Raina (2010), Loblay (2010), Abrol (2013), Mertia (2017), and Dahdah and Kumar (2018).
3 In fact, there also exist perversions to understanding technological outcomes through a social context. Arnold (2013) explains how race was often deployed to indicate how harmonious technological change would be (or would not be) to a society; and by extension a reflection of that society's modernity, or the necessity to forcefully inject modernity into that society which lacked it.

References

Abrol, D. (2013) 'New Science, Technology, and Innovation Policy: A Critical Assessment', *Economic and Political Weekly*, 48(9): 10–13.

Ahmad, A. (1983) 'Social Relations of Scientific and Technical Change', *Economic and Political Weekly*, 18(36/37): 1556–1561.

Arnold, D. (2013) *Everyday Technology: Machines and the Making of India's Modernity*, The University of Chicago Press, Chicago and London.

Banerjee, N., and Mitter, S. (1998) 'Women Making a Meaningful Choice: Technology and a New Economic Order', *Economic and Political Weekly*, 33(51): 3247–3256.

Bhalla, A.S. (1987) 'Can High Technology Help the Third World Take Off?', *Economic and Political Weekly*, 22(27): 1082–1086.

Bijker, W.E. (1997) *Of Bicycles, Bakelites, and Bulbs*, The MIT Press, Cambridge, MA.

Bijker, W.E. (2010) 'How Is Technology Made? That Is the Question!', *Cambridge Journal of Economics*, 34(1): 63–76.

Bijker, W.E., Hughes, T.P., and Pinch, T. (1987) *The Social Construction of Technological Systems*, The MIT Press, Cambridge, MA.

Bijker, W.E., and Pinch, T. (1987) 'The Social Construction of Facts and Artifacts: Or How the Sociology of Science and the Sociology of Technology Might Benefit Each Other', in Bijker, Hughes, and Pinch (eds.) *The Social Construction of Technological Systems*, The MIT Press, Cambridge, MA.

Bimber, B. (1990) 'Karl Marx and the Three Faces of Technological Determinism', *Social Studies of Science*, 20(2): 331–351.

Dahdah, M.A., and Kumar, A. (2018) 'Mobile Phone for Maternal Health in Rural Bihar: Reducing the Access Gap?', *Economic and Political Weekly*, 53(11): 50–57.

DN (2001) 'ICTs and Rural Poverty Alleviation', *Economic and Political Weekly*, 36(11).

Heilbroner, R.L. (1967) 'Do Machines Make History?', *Technology and Culture*, 8(3): 335–345.

Heilbroner, R.L. (1994) 'Technological Determinism Revisited', in Marx and Smith (eds.) *Does Technology Drive History: The Dilemma of Technological Determinism*, The MIT Press, Cambridge, MA and London, UK.

Hughes, T.P. (1969) 'Technological Momentum: Hydrogenation in Germany 1900–1933', *Past and Present*, 44: 106–132.

Hughes, T.P. (1994) 'Technological Momentum', in Marx and Smith (eds.) *Does Technology Drive History: The Dilemma of Technological Determinism*, The MIT Press, Cambridge, MA and London, UK.

Kalam, A.P.J.A. (2002) *Ignited Minds*, Penguin, New Delhi.

Kalam, A.P.J.A., and Pillai, A.S. (2004) *Envisioning an Empowered Nation: Technology for Societal Transformation*, Tata McGraw Hill, New Delhi.

Kalam, A.P.J.A., and Rajan, Y.S. (2002) *India 2020: A Vision for the New Millennium*, Penguin, New Delhi.

Kannan, L. (1994) 'Traditional Science and Technology: Learning from Legacy', *Economic and Political Weekly*, 29(24): 1487–1488.

Lerman, N.E., Oldenziel, R., and Mohun, A.P. (2003) *Gender & Technology*, The Johns Hopkins University Press, Baltimore and London.

Loblay, V. (2010) 'Experiments on Methodology in Reproductive Technology: Feminisms, Ethnographic Trajectories and Uncharted Discourse', *Economic and Political Weekly*, 44(44/45): 56–62.

MacKenzie, D., and Wajcman, J. (1999) *The Social Shaping of Technology*, Second Edition, Open University Press, Maidenhead and Philadelphia.

Marx, L. (1994) 'The Idea of Technology and Postmodern Pessimism', in Marx and Smith (eds.) *Does Technology Drive History: The Dilemma of Technological Determinism*, The MIT Press, Cambridge, MA and London, UK.

Marx, L. (1997) 'Technology: The Emergence of a Hazardous Concept', *Social Research*, 64(3): 965–988.

Marx, L. (2010) 'Technology: The Emergence of a Hazardous Concept', *Technology and Culture*, 51(3): 561–577.

Marx, L., and Smith, M.R. (1994) 'Introduction', in Marx and Smith (eds.) *Does Technology Drive History: The Dilemma of Technological Determinism*, The MIT Press, Cambridge, MA and London, UK.

Mertia, S. (2017) 'Timepass Development: Situating Social Media in Rural Rajasthan', *Economic and Political Weekly*, 52(47): 69–77.

Misa, T.J. (1994) 'Retrieving Sociotechnical Change from Technological Determinism', in Marx and Smith (eds.) *Does Technology Drive History: The Dilemma of Technological Determinism*, The MIT Press, Cambridge, MA and London, UK.

Mohan, D. (1995) 'Science Dynamics, De-Globalisation, and Social Developments', *Economic and Political Weekly*, 30(21): 1229–1234.

Morozov, E. (2013) *To Save Everything, Click Here: Technology, Solutionism, and the Urge to Fix Problems That Don't Exist*, Penguin, London.

NIELIT (2019) <nielit.gov.in>, National Institute of Electronics and Information Technology, Government of India, New Delhi.

NITI Aayog (2018) *National Strategy for Artificial Intelligence*, National Institution for Transforming India, Government of India, New Delhi.

Parthasarathi, A. (1969) 'Sociology of Science in Developing Countries', *Economic and Political Weekly*, 4(31): 1277–1280.

Prasad, C.S. (2005) 'Science and Technology in Civil Society: Innovation Trajectory of Spirulina Algal Technology', *Economic and Political Weekly*, 40(40): 4363–4372.

Raina, R.S. (2010) 'Situating Ethics in Technology and Science', *Economic and Political Weekly*, 45(23): 25–27.

Rajagopalan, C., and Singh, J. (1968) 'The Indian Institutes of Technology: Do They Contribute to Social Mobility?', *Economic and Political Weekly*, 3(14): 565–570.

Rath, S. (1994) 'Science and Technology: A Perspective for the Poor', *Economic and Political Weekly*, 29(45/46): 2916–2920.

Sarukkai, S. (2008) 'Culture of Technology and ICTs', in Saith, Vijayabaskar, and Gayathri (eds.) *ICTs and Indian Social Change: Diffusion, Poverty, and Governance*, Sage, New Delhi.

Scranton, P. (1994) 'Determinism and Indeterminacy in the History of Technology', in Marx and Smith (eds.) *Does Technology Drive History: The Dilemma of Technological Determinism*, The MIT Press, Cambridge, MA and London, UK.

Scranton, P. (1995) 'Determinism and Indeterminacy in the History of Technology', *Technology and Culture*, 36(2): S31–S53.

Sinha, P. (1967) 'Science and Society in India', *Economic and Political Weekly*, 22 April: 758–762.

Smith, M.L. (1994a) 'Recourse of Empire: Landscapes of Progress in Technological America', in Marx and Smith (eds.) *Does Technology Drive History: The Dilemma of Technological Determinism*, The MIT Press, Cambridge, MA and London, UK.

Smith, M.R. (1994b) 'Technological Determinism in American Culture', in Marx and Smith (eds.) *Does Technology Drive History: The Dilemma of Technological Determinism*, The MIT Press, Cambridge, MA and London, UK.

Swaminathan, P. (1991) 'Science and Technology for Women: A Critique of Policy', *Economic and Political Weekly*, 26(1/2): 59–63.

Upadhya, C. (2016) *Re-Engineering India: Work, Capital, and Class in an Offshore Economy*, Oxford University Press, New Delhi.

Upadhya, C., and Vasavi, A.R. (2006) 'Work, Culture, and Sociality in the Indian IT Industry: A Sociological Study', Report Submitted to the Indo-Dutch Programme for Alternatives in Development, National Institute of Advanced Study (NIAS), Bangalore.

Varma, R. (2001) 'People's Science Movements and Science Wars?', *Economic and Political Weekly*, 36(52): 4796–4802.

Vyasulu, V. (1976) 'Technology and Change in Underdeveloped Societies', *Economic and Political Weekly*, 11(35): M72–M80.

Wahyana, J. (1994) 'Women and Technological Change in Rural Industry: Tile Making in Java', *Economic and Political Weekly*, 29(18): WS19–WS33.

Winner, L. (1980) 'Do Artifacts Have Politics?', *Daedalus*, 109: 121–136.

2

A NETWORK STUDY OF TWO HANDLOOM WEAVERS' CLUSTERS[1]

This is the first study I employ to demonstrate how technological outcomes are deeply undergirded by sociological foundations. I present a network study of two handloom weaving clusters in Kerala, focusing on one community – the Saliyar community constituting one of these two clusters. The Saliyars were the predominant handloom textile weaving community for nearly a century, but they eventually declined about 20 years ago, and now occupy only a marginal position in the region. They dominated the industry not only in terms of market share and profits but also in terms of technique and design innovation in an industry where the very basis was, interestingly, the *constancy* of production technology. However, this techno-economic dominance was not solely business acumen but greatly on account of their social cohesion – as was their decline. In other words, I demonstrate in this chapter how the techno-economic decline of this community was fuelled by significantly higher social cohesion in its networks, compared to the other handloom-engaged communities in these two clusters.

New and valuable information on technique and design innovation in this milieu are created and shared not by formal research and development in corporations but through interesting dynamics that we might term "information sharing through informal interaction" among individuals and groups. These, naturally, are therefore contingent on complex social relations. When these dynamics in an industrial cluster, such as the handloom clusters here, are influenced by complex social relations for an extended period of time, the emergent path that the cluster takes is noteworthy. The experience of the Saliyar community cluster in Balaramapuram town (in Trivandrum District, in the southernmost tip of Kerala state), whose hereditary occupation is handloom textile production, is notable in this regard. While even 30 years ago every Saliyar household was a weaving and producing unit, at the forefront of the sector, today there are less than a handful of Saliyar weavers in this cluster. As stated previously, I attribute this to a heavy involvement of social relations in business and production relations.

Social relationships enter economic relations at almost every stage of economic activity, from the selection of economic goals to the organisation of relevant resources (Portes, 1995: 3). "Social embeddedness," or just "embeddedness," a central concept in economic sociology, implies that economic behaviour and decision making reside greatly within social relations (Granovetter, 1985; Woolcock and Narayan, 2000; Krippner, 2001). Embeddedness as a concept is based on the same notions as social capital: the un-atomised nature of the individual, the weakness of the "immediate utility" approach in explaining social relations, the logics underlying the formation of institutions and norms, and the fact that these cannot be removed from the social, cultural, and cognitive contexts and identities they are implanted in (Boschma, 2005; Ghezzi and Mingione, 2007). The nature of information sharing in low-technology (or for that matter, even high-tech) environments is often contingent upon the social identities of the economic agents providing or receiving it (as work by Brian Uzzi [1996, 1997] and others has illustrated), to the extent that social identities and affinities between social groups may even decide whether the innovative information is shared at all. Production, exchange, business, and technological relations are not peripheral to – and may develop as emergent properties of – complex social relations in the region, and may be characterised by homophily, which is the tendency of agents to be linked to other agents with similar characteristics (Jackson, 2008), i.e., the tendency of individuals to associate disproportionately with those similar to themselves (Golub and Jackson, 2009, 2012). Homophily and effective communication breed each other; individuals departing from homophily often face hindrances due to differences in social status, beliefs, language, and so on, which may distort meanings of messages (Rogers, 2003).

The main proposition in this study is that high homophily, high insularity, and excessive involvement of social relations in business relations led the Saliyars' networks to be too cohesive, compared to the networks of the other communities in the region, and restricted the Saliyars' participation in information exchanges regarding innovative and valuable information. We study production, information, and social networks of the Saliyar community cluster at Balaramapuram and compare them to the networks of various other communities in a socially heterogeneous cluster called Payattuvila, near Balaramapuram, producing exactly the same product but occupying a more dominant position in the industry. As mentioned at the outset, this study is placed in an interesting technological setting characterised by technological *constancy*. Weavers in the numerous handloom clusters across Balaramapuram town specialise in weaving of cotton textiles in a style unique to Kerala, where the antiquity of handloom technology and product design are the very *basis* of its consumer demand and its niche in domestic and international markets. Upgrading the technology to electric "powerloom," and changes in design deviating from a traditional standard, would in fact *endanger* the industry and consumer demand. However, this constraint does not imply that there are no avenues for the creation and diffusion of new information. On the contrary, it may be even more imperative to share new information given this atypical situation of knowledge improvements constrained by unchanging production technology.

The chapter first introduces the handloom industry in Balaramapuram and the Saliyars. It then moves to explaining the field procedures and survey methods.

27

In the next section, I begin comparing the networks of the Saliyar community cluster and the networks of the other communities in terms of their homophily and embeddedness, and proceed to analyse them.

Overview of handloom at Balaramapuram, and the Saliyar cluster

The Indian textile industry as a whole contributes to around 2% of GDP, to 15% of export earning, and employs a staggering 45 million people directly and indirectly; the handloom sector within textiles produces 15% of cloth and employs 4 million people (MoT, 2018). In Kerala state, handloom is the second-largest traditional industry, employing nearly 100,000 people, concentrated in Trivandrum district in the south and Kannur district in the north (GoK, 2018). Many operate under a cooperative system, while the rest operate either independently or under master-weavers.[2] Procurement and marketing are undertaken primarily by state agencies such as Hantex (Kerala State Handloom Weaver's Co-operative Society) and Hanveev (Kerala State Handloom Development Corporation). While handloom in most of Kerala concentrates on household products, clusters in Trivandrum district (including in Balaramapuram) have always had a niche in traditional attire for many decades (Rajagopalan, 1986). The Geographical Indication Tag with Intellectual Property protection for ten years was granted to the "Balaramapuram *sari*" in January 2010 by the Government of India (MoT, 2010).[3] In this region, the Fly-Shuttle loom and the Pit loom are the most popular technologies, preferred over other weaving technologies such as the Dobby or Jacquard looms (Hanveev, 2006). Yet, the unchanging nature of production technology appearing as a constraint has not deterred weavers in this region, who have displayed remarkable resilience, having adjusted to changing market requirements with low energy-intensive and low capital-cost production methods (Niranjana and Vinayan, 2001).

Cooperative organisation in production was promoted throughout India, and especially in Kerala given its long communist rule. Cooperative societies dominate 98% of the handloom industry in Kerala (GoK, 2018). But in reality, most registered cooperatives are said to exist only on paper; even in 2001, at least 250 out of the 366 listed cooperative societies in Trivandrum district were said to be either nonexistent or nonfunctional (Niranjana and Vinayan, 2001). Weaving in this district, including in Balaramapuram town, is hence done mostly at the weaver's residence, or under a master-weaver's unit (in Kannur district, however, cooperatives are still dominant). Across most of India too, handloom is generally household based, with production shared by the whole family and not only the weaver at the loom (Raman, 2010). According to the Kanago Committee report, on average around 55% of a weaver's family are gainfully employed at various stages of production (Niranjana and Vinayan, 2001; Soundarapandian, 2002). Besides household units, there are also a small but significant number of factories and large units in Kerala.

Handloom weaving in Trivandrum district is centuries old. The early 1800s were a turning point when the Maharaja of erstwhile Travancore built up a weavers' cluster of various communities at Balaramapuram town (Hanveev, 2006).[4] The late 1800s saw another significant turning point when the then Maharaja brought in, from the neighbouring state, weaver families belonging to five particular weaving communities, the most prominent among these being the Tamil-speaking Saliyars, all settled as a cluster on one set of streets in central Balaramapuram (Niranjana and Vinayan, 2001). The Saliyar community cluster of weavers in Balaramapuram town in Trivandrum district was, and still is, surrounded by socially heterogeneous (and predominantly Malayalam-speaking) clusters of other weaving communities. The Saliyars long were the dominant weaving community in the region, catering to orders from the royal family and reportedly introverted for business and social relations but not out of any animosity towards other communities. Over time, the dominance of the Saliyars as handloom producers began decaying, gradually reducing into a rather marginal role. There is no single dominant weaver community in Balaramapuram today, but the position of the Saliyars in weaving has eroded and this community is now almost entirely involved in activities at other stages of handloom clothing production, from the same location.

When the research for this study was undertaken in 2011–2012, the population of the Saliyar community in Balaramapuram, according to the local-government sources, was almost 1,000, residing in over 300 households in Ward 7 and 8 in south-central Balaramapuram town. But the number of Saliyar households in this cluster who deal with handloom production today is just under 30, and most households now participate in pre-weaving activities such as plying and yarn supply. A few members of the oldest generation, above 75 years of age, have recently retired altogether. A temple is centrally located in the community region, neighboured by a community hall, as well as a now-defunct Saliyar community welfare organisation and cooperative society. The Saliyar Cluster is composed of four main roads (Single Street, Double Street, New Street, and Vinayagar Street) about 30 feet wide and each 500 feet long, radiating out from the four walls of the temple, with numerous small alleys in between. While until even 40 years ago the Saliyar Cluster had a pit loom at each house, there are today barely any households with regularly functioning looms, undertaking weaving as a professional activity. In the past, every family member was involved in some stage of production – women in weaving, and men in the many other activities or weaving only large textiles – but today this handful of households are operated mostly by male weavers, including two master-weavers under whom a few looms are operated by employees from other communities. Old business and family links with Valliyur, near Nagercoil town (a town in southern Tamil Nadu state, around 50 kilometres from Balaramapuram, from where most Saliyars in Balaramapuram trace their ancestry), are still maintained by the Saliyars, as are links with small producers in Surat (a major town in Maharashtra state in western India) for gold thread, and with Muslim beamers in Trivandrum district for ordinary yarn. It may be noted that there are a few agents (mainly shop owners) who operate within the Saliyar community cluster but do not belong to the Saliyar community.

Field procedures and questionnaire

To study the cohesiveness and eventual decline of this community, I studied the networks of Saliyar community members who are engaged in any stage of handloom production (not the entire community residing in the Saliyar neigh-bourhood). Following Arora (2009), I mapped out the social, production, and information networks of the Saliyars, paying close attention to the expected homophily and cohesiveness of the community. I compared these networks to the expected non-homophilous and non-cohesive links of a similar handloom textile producing cluster of a similar (though socially heterogeneous) popula-tion, covering a similar geographic area as the Saliyar community, in a village called Payattuvila in the vicinity of Balaramapuram. Networks in both cases extend outside the cluster, and to suppliers and procurers besides Hanveev and Hantex.

This study engaged with seven individuals to first establish awareness and familiarity of the weaver clusters at Balaramapuram – three in government, one large retailer in Balaramapuram, one prominent (non-Saliyar) master-weaver in Balaramapuram town under whom almost 25 looms operate, one elderly Saliyar weaver, and another elderly master-weaver at Payattuvila. Sessions with these respondents in May and September 2010 gave a broad picture of the Saliyar com-munity and social relations in general in the region's handloom clusters.

The unit of analysis within the community was the household unit engaged in handloom production activity, and the targeted interviewee in each household was the member of the family currently engaged in handloom textile production at any stage.

Two community elders in the Saliyar community and one elder at Payattuvila served as key informants in the respective clusters, who provided us access to the households. An interpreter's support was inevitable for the entire study, as all enquiries had to be undertaken in Malayalam.[5] After a pilot visit in early Sep-tember 2010, interviews of household units in the Saliyar community at Balara-mapuram and at the Payattuvila cluster were conducted across 2011. This study conducted interviews with all households in the manner of conversation but also relied on a sectioned questionnaire. The questionnaire for network structure sur-vey, inspired in structure by Arora (2009), was sliced into five modules – basic information on household and production activity, professional network, social network, information network, and miscellaneous information.

1   The Basic Information section elicited the main activity of the household, members involved in the main activity, and whether this was the only income-generating activity, as well as the capital employed and the technol-ogy used for the respective production activity.

2   The Professional Network section elicited lists of main providers of input/raw material, consumers, and financiers.

3   The Social Network section elicited lists of relatives and friends who were *very* close, such that they met the respondent at least once a day, and who the respondent approached first in the event of domestic and family emergencies.

4   The Information Network section elicited lists of the first individuals or agencies the respondent would approach if there were any business or production issues, such as new consumers, new market trends, new technologies (in activities besides weaving like plying, dyeing, etc.), new designs, tastes, etc., i.e., any new piece of news on production and commercial know-how and on tastes and preferences. The section also enquired about the method of communication used with the above individuals and agencies. When naming individuals or other agents, it was asked not whom a respondent simply "knew" as, say, a provider of the latest know-how but rather who the *first* few individuals or agents the respondent would approach when curious on the latest developments in products, production processes, designs, trends, etc.

5   The Miscellaneous section enquired about whether there was anyone such as a "most influential" person in the cluster and what the nature of relations and reputation kept up with the respondent was. It also enquired about the role of the state and local handloom and textile cooperatives – whether they were of any use at all for providing new information.

It was essential for the respondent to name his or her social group, activity, and exact location, as a part of the Basic Information section. In the case of the Saliyars, no restriction was placed to the respondent to preferably list individuals in the same community; on the contrary, enquiries were also specifically made on whether there were non-Saliyar actors in their networks. Understandably, a complete list of actors in the Saliyar community cluster or Payattuvila cluster networks was not possible to acquire, hence I relied upon a snowballing listing of respondents for interview. Free choice was adopted (as opposed to fixed choice, where respondents are told how many individuals to list in their network). During pilot interviews with the first few households, I had attempted to rank individuals in terms of importance, but that effort was abandoned eventually. Given the fact that the handloom textile profession at the household level requires meeting clients and suppliers on a daily basis, this survey's enquiry on strength of ties was limited.

There is little distinction between "home" and "production area" in households in both the Saliyar and Payattuvila clusters because a major part of the house, sometimes the very entrance, is used for production along with other domestic purposes. In almost all households surveyed, handloom production is undertaken almost every day and intensified only during seasons such as weddings, when custom-made orders are highly demanded. There are no definite "workdays" (almost every day is a workday), no rigid work routines except at the master-weavers' units, and production at almost all stages uses traditional technologies – hand-plied yarn for plying, pit-looms and fly-shuttle looms for weaving, and so on – except at the spinning stage, where electrically operated spinning machines are employed, located inside the house.

Network analysis

Besides the Saliyars, I identified three other communities (categorised along lines of caste) in the network, denoted with anonymity here as Community-II, Community-III, and Community-IV. Actors in the network whom I could not categorise by community – including state agencies, showrooms, shops, financiers, and media sources – were categorised as Community-NIL. Many actors operate in locations beyond the two clusters, spread across Kerala state and India. Table 2.1 summarises the community and regional distribution of the 62 actors in this network.

The networks

I employed UCINET (Borgatti et al., 2011) to generate the network diagrams. In all three networks, I grouped actors by location, and differentiated their community by shape. Hence, for example, node 42 in the network diagrams that follow is a square (Community-II member) operating in the Payattuvila Cluster; node 47 is a down-triangle (Community-IV member) operating in Kerala; node 14 is a circle (Saliyar community member) operating in the Saliyar Cluster; and so on. Occupations – weavers, yarn sellers, plyers, etc. – are not assigned attributes, as it might result in some confusion in the diagrams to have nodes classified into a large number of occupational categories. Table 2.2 shows the occupational distribution of actors in each cluster.

"Raw Input Supplier," "Miscellaneous Customer," "Media Sources," and "Others" have been excluded from Table 2.2, as they operate beyond the two clusters.

In the Professional Network (Figure 2.1), it can be seen that nodes in the Saliyar Cluster and Payattuvila Cluster may be heavily connected within, but there are also a number of dyads that span the two clusters: (13,33), (13,43), (13,41),

Table 2.1 Distribution of actors by community and region

	Saliyar Cluster	Payattuvila Cluster	Balaramapuram	Kerala state	Tamil Nadu	Rest of India	Total
Saliyar	17	0	0	0	2	0	19
Community-II	0	7	0	0	0	0	7
Community-III	0	3	0	2	0	0	5
Community-IV	2	4	0	0	0	0	6
Community-NIL	0	3	9	5	2	6	25
Total	19	17	9	7	4	6	62

Source: Author's fieldwork

Table 2.2 Occupational distribution of actors in each cluster

	Weaver	Retail and wholesale shop	Plyer	Yarn shop	Spinner	Master weaver	Financier	Cooperative	Total
Saliyar Cluster	3	8	4	2	1	1	0	0	19
Payattuvila Cluster	10	0	0	0	0	4	1	2	17

Source: Author's fieldwork

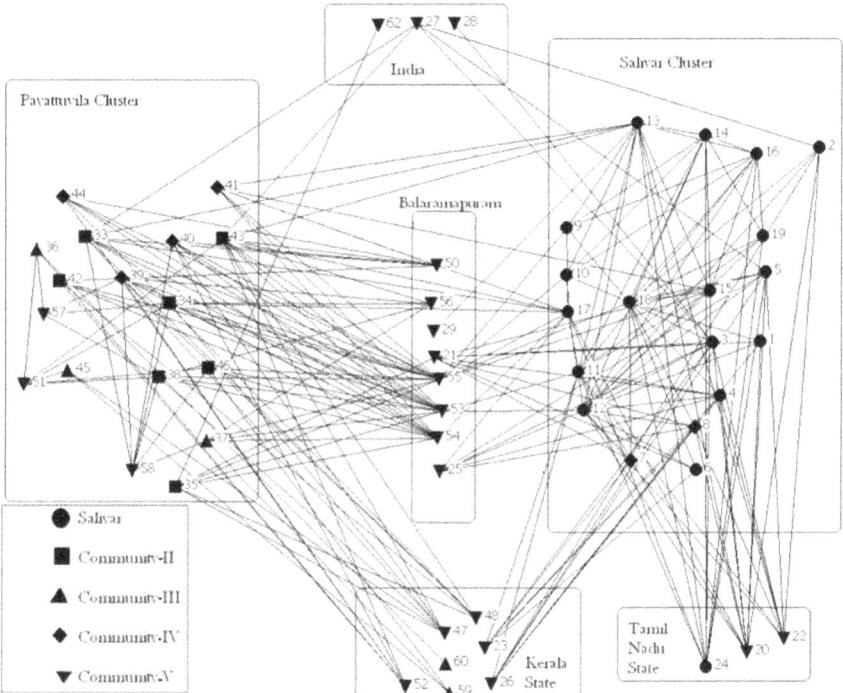

Figure 2.1 Professional network
Source: Kamath and Cowan (2015)

(15,41), (17,33), (17,41), (17,43). Besides these dyads, there are a few nodes that bridge both clusters, including 21 (retail shops), 53 (plyers), 54 (yarn sellers), and 55 (yarn spinners), all in Balaramapuram. In the Information Network (Figure 2.2), too, a number of dyads span the two clusters: (11,46), (13,41), (17,33), (17,38), (17,39), (17,40), (17,42). Besides these dyads, nodes that bridge both clusters include 31, 47, and 48 (large retail sellers that operate across various

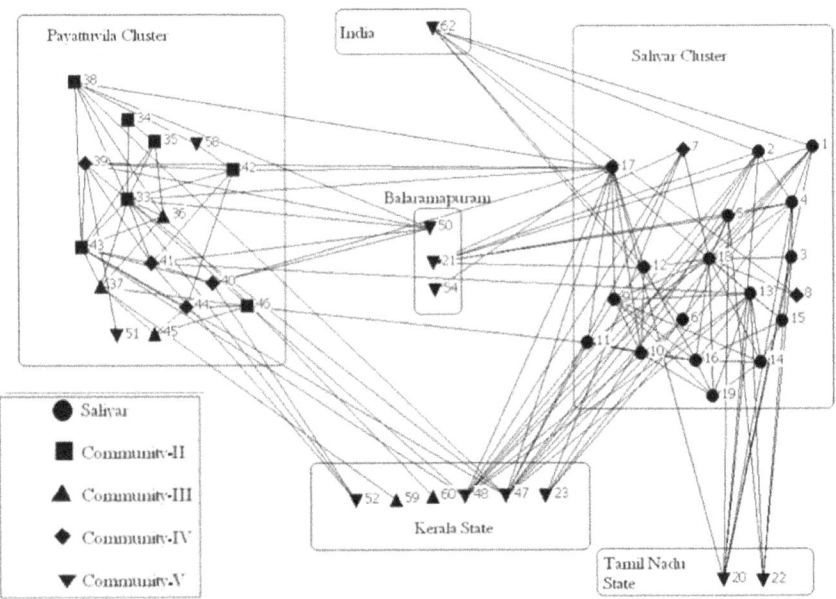

Figure 2.2 Information network
Source: Kamath and Cowan (2015)

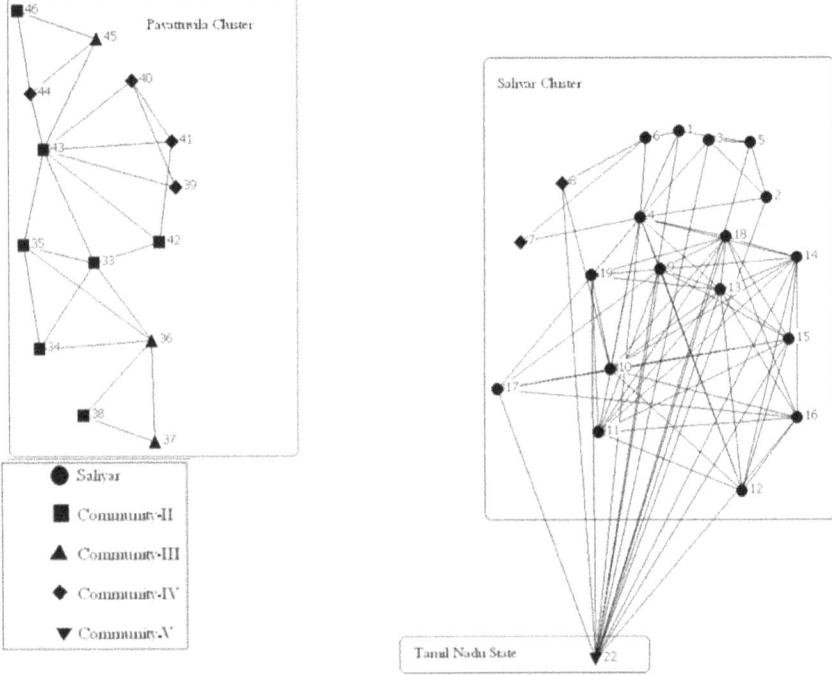

Figure 2.3 Social network
Source: Kamath and Cowan (2015)

districts in Kerala and India). Actors 47 and 48 are two interesting actors, who will be revisited in a later section.

Evidently, the Social Network (Figure 2.3) is completely polarised along cluster lines. Here, the Payattuvila Cluster's community heterogeneity is particularly noticeable. Node 22 in the Saliyars' social network stands out, this node referring to a few families in Nagercoil town whose inhabitants are Tamil-speaking, and some of whom are even distantly related to the Saliyars. Members of the two clusters know each other by name, but as explained earlier, the questionnaire requested names of relatives and friends who were *very* close.

Descriptive observations on homophily and links across regions

I now move to describing homophily exhibited by the four communities, to demonstrate the nature of affinities of actors to other actors in their own community, in each of their networks. I also look at the geographic spread of each community's links. Most measures of homophily are closely related to the in-group/out-group ratio by Duysters and Lemmens (2003), based on Wasserman and Faust (1994).[6] Generally, homophily H_i of an actor i can be measured as the ratio of the number of links of actor i to the community she belongs (s_i), to the total number of links she possesses ($s_i + d_i$), that includes links with other communities (d_i).

$$H_i = \frac{s_i}{s_i + d_i} \qquad [2.1]$$

This measure, cautioned by Coleman (1958) and Currarini et al. (2009), fails to account for group size. Instead, I use Inbreeding Homophily (*IH*) in Currarini et al. (2009), which normalises to control for the sizes of the different groups in the network.

$$IH_i = \frac{H_i - W_i}{1 - W_i} \qquad [2.2]$$

H_i is the basic homophily measure in [2.2], while W_i is the size of the group that actor i belongs to. An average of the *IH* of all members in the group A that actor i belongs to, gives the homophily of the whole group IH_A. Group A is said to display Complete Inbreeding when $IH_A = 1$, Pure Baseline Homophily when $IH_A = 0$, Inbreeding Homophily when $IH_A > 0$, and Inbreeding Heterophily when $IH_A < 0$. We measure homophily using *IH* for each community (except Community-NIL, of course), and across professional, information, and social networks. As can be seen in Table 2.3, the descriptive results of these measures provide evidence on the relatively high community cohesion of the Saliyars.

It can easily be seen that the Saliyars are highly homophilous in all three networks. Only the Saliyars and Community-III show Professional-Network homophily; the other two communities display heterophily in this network. In

Table 2.3 Homophily across communities

	Professional network	Information network	Social network
Saliyars	0.500	0.526	0.930
Community-II	−0.145	0.310	0.388
Community-III	0.178	0.330	0.057
Community-IV	−0.120	−0.009	0.160

Source: Author's computations, based on fieldwork

the Information Network, the Saliyars show much higher homophily than do Community-II and Community-III (but not Community-IV, which has a very small population). The Saliyars also show the highest homophily in their Social Network, by virtue of their members residing in a socially homogenous cluster. A Welch t-test to assess differences between means between the Saliyars and the other three communities with respect to these results found that while differences were mostly significant in the professional and social networks, they were very insignificant in the *information networks* (except between the Saliyars and Community-IV).

It trends out that there appears to be non-significance of differences in Information Network homophily. This requires attention, as a central idea in this study is that access to *information* has always been vital to prosperity in this industry. A lack of significant differences in Information Network homophily creates a problem for the core argument of this study that rests deeply on homophily. However, we return to the analysis of homophily in a short while to re-check the perspective of analysis.

For now, to continue, let us also study how each community differs in the geographical spread of their professional and information links (Tables 2.4 and 2.5).

It can be seen in Table 2.4 that the Saliyars are the only community maintaining more within-cluster professional links compared to out-of-cluster links. But when we disaggregate the out-of-cluster professional links to explore this, we see that the Saliyars keep around 42% of their out-of-cluster professional links with Tamil Nadu (particularly with Nagercoil town), i.e., to their own community members living there. Hence, the Saliyars' out-of-cluster links are not exactly as heterophilous as they might initially suggest. The Saliyars appear to prefer homophilous connections even in their out-of-cluster links.

In their information links, all communities maintain more within-cluster than out-of-cluster links, as seen in Table 2.5, of which Community-IV maintains the highest out-of-cluster information links. Interestingly, the Saliyars seem to maintain the lowest within-cluster information links too. But this is not surprising, as on disaggregating the out-of-cluster information links, we observe that around one-fifth of the Saliyars' out-of-cluster information links are, again, with Tamil Nadu. Once again, the Saliyars have linked out-of-cluster for new information with their own community members living in Nagercoil town.

Table 2.4 Geographical distribution of professional links

Community	Proportion of links within cluster (%)	Proportion of links outside cluster (%)	Disaggregated out-of-cluster Professional links (% of professional links outside cluster)			
			Balaramapuram town	Kerala state	Tamil Nadu	Rest of India
Saliyars	61	39	30.70	20.85	42.54	5.91
Community-II	16	83	67.28	28.14	0	4.58
Community-III	33	67	33.33	66.67	0	0
Community-IV	19	81	64.19	30.24	5.57	0

Source: Author's computations, based on fieldwork

Table 2.5 Geographical distribution of information links

Community	Proportion of links within cluster (%)	Proportion of links outside cluster (%)	Disaggregated out-of-cluster Information links (% of professional links outside cluster)			
			Balaramapuram Town	Kerala State	Tamil Nadu	Rest of India
Saliyars	56.1	43.9	13.11	40.22	21.76	24.91
Community-II	79.3	20.7	32.17	61.72	0	6.11
Community-III	61.9	38.1	49.99	49.99	0	0
Community-IV	91.7	8.3	0	66.65	0	33.35

Source: Author's computations, based on fieldwork

Overall, we have seen that the Saliyars are significantly more homophilous than the other communities in terms of professional and social networks but not in terms of the information network. They seem to exhibit an interesting homophily even in the geographical spread of their links, towards their own community in the neighbouring state.

Embeddedness and path lengths to influential information actors

To continue the network analysis, we must study the extent to which a community's production and information networks are embedded in its social network. This, along with the previous section on homophily, seeks to address the social cohesion of the Saliyars in a more holistic manner. To anticipate the results of this study, the Saliyars appear to be much more cohesive than other groups, prompting the argument that their excessive social cohesion caused their decline.

Before enquiring about the magnitude of embeddedness of each community, we must perform a correlation to test the association of the social network with the professional and information networks individually. The test of the extent to which one network resembles another is a correlation between the adjacency matrices of these networks. This test is required because calculating confidence bounds on a correlation of adjacency matrices is not straightforward, due to the interdependence both among cells of the matrix and among different properties of the nodes (Krackhardt, 1987; Arora, 2009). A standard technique in network analysis for generating confidence intervals is the Quadratic Assignment Procedure (QAP). QAP is a permutation text which in essence generates a distribution of a statistic according to some null hypothesis, by permuting rows and columns (simultaneously) of one of the matrices, consistent with the null hypothesis. This creates a frequency distribution of the test statistic, which is taken as an estimate of the true underlying distribution. This permits estimation of confidence intervals around the observed statistic. Results of the QAP test on the correlations of the social network with the professional and business networks are given in Table 2.6.[7]

Table 2.6 indicating the Pearson Correlation shows the observed value of correlation between the social network matrix and the other two network matrices individually, 0.195 and 0.397. The Average Random Correlation is simply the mean value of the correlations between the professional (information) network and each of the permuted social networks. The number of random correlations that were larger than 0.195 and 0.397 was zero, as shown in the "Percentage (Larger)" column, i.e., none of the 5,000 random permutations produced a correlation higher than 0.195 or 0.397 for the respective matrices. With these results, we can say that the correlation between the social-network matrix and the professional-network and information-network matrices, respectively, is statistically significant, and the correlations between these matrices are unlikely to have occurred by chance. With this, let us now define embeddedness directly. Here we follow Arora (2009: 154).

$$Social\ Embeddedness\ of\ Production\ Network = \frac{\sum_{ij} P_{ij} S_{ij}}{\sum_{ij} P_{ij}} \qquad [2.3]$$

$$Social\ Embeddedness\ of\ Information\ Network = \frac{\sum_{ij} K_{ij} S_{ij}}{\sum_{ij} K_{ij}} \qquad [2.4]$$

P_{ij}, K_{ij} and S_{ij} are the adjacency matrices of the production network, information network, and social network respectively. An element in P_{ij} (equivalently K_{ij} and S_{ij}) takes the value 1 if i and j are professionally (or for information or socially) connected, and 0 otherwise. That is, if two actors are connected both by a production link as well as a social link, their professional connection is said to be socially embedded, and likewise for their information network link. Table 2.7 shows these measures of embeddedness for the different communities.

What is striking, and apparently discordant with the observations on homophily, is that the Saliyars are relatively *less* information-network embedded. In

Table 2.6 QAP correlation results

	Pearson Correlation	Significance	Average random correlation	s.d.	Percentage (larger)
Professional Network Matrix	0.195	0.000	0.000	0.025	0.000
Information Network Matrix	0.397	0.000	0.000	0.026	0.000

Source: Author's computations, based on fieldwork

Table 2.7 Social embeddedness of each community

	Professional network	Information network
Saliyars	0.379	0.456
Community-II	0	0.544
Community-III	0	0.500
Community-IV	0.456	0.278

Source: Author's computations, based on fieldwork

addition, the differences are not statistically significant. These results are discordant with the initial conjecture that community cohesiveness is an intrinsic characteristic of the Saliyars, which is at the root of their techno-economic decline. In addition, the embeddedness statistics give a mixed message: professionally the Saliyars seem to operate within their social network, but when it comes to accessing information, they appear less embedded than two other communities.

Let us quickly visit one more network measure. It is known from the literature that information flows between randomly selected pairs of agents may be much less important than flows from particular agents, as short path lengths to more well-informed actors or groups are known to be conducive to fast and effective diffusion of information in a network (Cowan, 2004). We can estimate this for a group by the proximity of each community to nodes in the network that are influential trendsetters in the industry, what I term "influential information actors" (IIAs). There are two IIAs in this network: two successful upmarket retail showrooms, actors 47 and 48, recognised by industry and the state government as being at the forefront and cutting edge of design and innovation in the handloom industry in Kerala (GoK, 2007: 21). These were also mentioned by the respondents (recorded in the Miscellaneous section of the questionnaire) as influential trendsetters on design and market information on handloom in Kerala. To calculate path length, I relied on a simple measure of geodesic distance, in this case the distance from an actor to nodes 47 and 48. We can compute a simple mean of the distance to both these nodes for each actor in a group. This individual-actor measure can be averaged across all actors in the group to obtain the mean path length of the group to the IIA (Table 2.8).

Table 2.8 Mean path length to influential information actors (IIAs) of each community

	Mean path length to IIA
Saliyars	1.68
Community-II	1.71
Community-III	2.33
Community-IV	2.08

Source: Author's computations, based on fieldwork

According to these results, the Saliyars on average appear relatively closer to the IIA, and should therefore have had easier access to new information, providing a competitive advantage and supporting their long-term viability. The impression conveyed by these results is puzzling, as the techno-economic story on the ground has been different. We should also observe, though, that differences in means between communities with respect to these results were found to be not significant.

Neither homophily nor embeddedness

The literature makes a strong case that any group exhibiting either strong homophily or strong embeddedness, and especially if exhibiting both, would in time find itself at a competitive disadvantage (see Kamath [2015] for a detailed treatment of this theme). The networks of the Saliyar community show homophily, but their professional and information networks are not strongly embedded in their social network. At least, they are not too significantly different in terms of embeddedness, compared to those groups whose presence in weaving has not declined. Further, the Saliyars in general are closer, in network terms, to the most important information actors than are the other groups. Thus, in sum, this appears to suggest that the Saliyars have networks that are not particularly socially embedded, are close to the important information sources but show some homophily. The simple conclusion to draw would be that only homophily matters. But this seems too quick, particularly because information about technologies and markets must surely be central for survival in any business. I argue, in the following section, that treating homophily and embeddedness jointly provides a more complete picture of "social cohesion" as such, and lends a satisfactory consistency with the experience in reality of the gradual decline of the Saliyar position.

A measure of joint cohesion

The Saliyars have networks that are not particularly embedded, are close to important information sources but show some homophily. None of these differences between groups is statistically significant, however. Neither of the conventional indicators of cohesion shows results indicating significant differences between the Saliyars' networks and the others' networks, which is not consistent with the fact that the historical trajectories of the different groups are very different. The statistics even seemed to suggest that the Saliyars had more open Information Networks. This suggests that the existing measures employed so far may be too simplistic, and that a joint measure may lead to an improvement in the way we think of or measure cohesion. I argue here that treating homophily and embeddedness *jointly* can provide a more comprehensive picture of cohesion. It may even provide an improved explanation of the deep-seated differences between the Saliyars and the others in lived experience.

An agent's production (or information) link may be overlapped by her social link (i.e., it may be embedded), but this need not be homophilous. Similarly, her homophilous production (or information) link may not be socially embedded. I therefore offer four kinds of links in the cohesion spectrum:

1 **Non-Homophilous Non-Embedded,** when an agent's link is neither homophilous nor socially embedded.
2 **Embedded but Non-Homophilous,** when an agent's socially embedded link is non-homophilous.
3 **Homophilous but Non-Embedded,** when agent's homophilous link is not socially embedded.
4 **Homophilous-Embedded,** when an agent's socially embedded link is also homophilous.

We would expect that *homophilous-embedded* links are more detrimental in the long run than the other three types, because they draw the combined deteriorative effects of both embeddedness and homophily, while the others may bring in deteriorative effects of only embeddedness or only homophily. It follows that one must beware not only over-embeddedness but also its *combined effect* with homophily. The literature on embeddedness (see Kamath, 2015) has not entirely ignored the fact that it is not monolithic but has not articulated it very clearly either. This study contributes to the disentangling of social embeddedness by demonstrating through the case of the Saliyars, the vastly different effects of homophilous-embeddedness and non-homophilous-embeddedness, from merely "embeddedness" or "homophily." The Saliyars may have declined not simply because they possessed homophilous links or embedded links but due to the effect of their predominantly homophilous-embeddedness networks. Let us measure this cohesion, by calculating an actor's homophilous-embedded links as a proportion of her total links to calculate the extent of her homophilous-embeddedness.

Tables 2.9 and 2.10 show a different picture of each community's cohesion, and compared with Table 2.7 there is significant change. In their professional and information networks, the Saliyars show a high proportion of links of the most cohesive type: homophilous-embedded. They show essentially no embedded non-homophilous links but a relatively high proportion of non-embedded homophilous links, in both networks. The strongest pattern in these tables is that

Table 2.9 Proportion of cohesive links in professional network (in %)

	Non-embedded and non-homophilous	Non-embedded but homophilous	Embedded but non-homophilous	Homophilous-embedded
Saliyars	30.10	31.67	0.66	37.24
Community-II	100	0	0	0
Community-III	51.67	48.33	0	0
Community-IV	97.90	0	2.08	0

Source: Author's computations, based on fieldwork

Table 2.10 Proportion of cohesive links in information network (in %)

	Non-embedded and non-homophilous	Non-embedded but homophilous	Embedded but non-homophilous	Homophilous-embedded
Saliyars	23.68	30.25	0	46.07
Community-II	39.35	6.22	20.85	33.57
Community-III	22.22	27.78	38.89	11.11
Community-IV	67.74	4.46	22.38	5.42

Source: Author's computations, based on fieldwork

the Saliyars have (statistically) significantly more links of the most cohesive type, in both their professional and their information networks. For the Saliyars, the most common type of link they display is the most cohesive type, and compared to the other groups, the Saliyars appear to have the most strongly cohesive links. These results are concordant with our initial proposition that the Saliyars' networks, particularly their information networks, are very cohesive.

The new measure provides a better way of distinguishing between the different communities in terms of their cohesion, and may be much more effective, eventually, in explaining how and why social relations affect techno-economic outcomes. Neither of the conventional indicators of cohesion – homophily or embeddedness – found strong differences between the communities' networks. In fact, results computed using the conventional embeddedness and homophily measures were, by and large, even statistically insignificant regarding the differences between the Saliyars' networks and the others' networks. Hence, though the Saliyars have had a very different history than the other communities, traditional measures of cohesion conveyed the impression that there was little difference between the Saliyars' links and the others' links. We see nothing in the results generated from these conventional measures that would explain the difference in experiences and histories of the different groups. By contrast, Tables 2.9 and 2.10 show results from using more sophisticated measures of cohesion that capture both embeddedness and homophily. The joint measures identify statistically significant differences between the Saliyars and other communities, which are consistent with the proposition that the Saliyars' heavily cohesive networks – particularly their information network – are connected to their eventual decline.[8]

The conventional measures of cohesion therefore may have been simplistic, and they also did not find any difference between the Saliyars' networks and the others' networks, when there evidently are strong differences that meet the eye when one visits these clusters. Both embeddedness and homophily matter, and the new measure indicates the substance of this joint effect as it brings out significant differences between the communities' networks. The dominance of the Saliyars' homophilous-embedded links, significantly different when compared to the others, indicates the kind of network characteristics that would explain the different histories, and hence the different techno-economic outcomes. Socially cohesive links are detrimental under certain conditions, a position that applies well to the Saliyars, and undergird technological and economic experiences.

The final task is to test whether there is a significant difference in the proportion of homophilous-embedded links between weavers and non-weavers, in these clusters, in their professional and information networks. For this, one needs to group actors, across all communities in the two clusters, into Weavers and Non-Weavers. The homophilous-embeddedness in each group's professional and information network links needs to be computed. It was expected that, in general, weavers in these handloom clusters would have *lower* homophilous-embeddedness in their networks than non-weavers, as weaving is an activity that intensively involves information gathering from various sources, sustainable only by those actors who are less cohesive. Non-weavers, on the other hand, can afford to be more homophilous-embedded in their professional and information networks.

These expectations were affirmed with the results in Table 2.11. Weavers appear to be significantly less homophilous-embedded in their professional and information networks compared to non-weavers. And as Saliyars comprise mostly non-weavers (while the other communities in general, and Payattuvila Cluster as a whole, comprise mostly of weavers), this result stands in concordance with the results in Tables 2.9 and 2.10.

With these results, we can ascertain that it is indeed predominantly their homophilous-embedded networks, not just their embeddedness or homophily alone, that captures the Saliyars' cohesion, and we may employ these results to understand their techno-economic decline. Homophily and embeddedness do matter, but the combination effect brings out the true differences between the communities.

Table 2.11 Proportion of homophilous-embedded links between weavers and non-weavers (in %)

	Professional network	Information network
Weavers	11.43	25.09
Non-weavers	24.82	37.50

Source: Author's computations, based on fieldwork

Conclusions

The idea of complex social relations shaping business and production relations between economic agents is not new, and these relations are often characterised by embeddedness and homophily. In this chapter, I have presented the case of the Saliyars of Balaramapuram, who exhibited cohesion to such an extent that it has marginalised their community cluster in the overall scheme of production and information sharing relations in the Balaramapuram handloom clusters. Lessons drawn from this cluster's experience show that social embeddedness in combination with thick homophily in production and information networks can have influence exclusion in information networks (which are in a large sense "techno-economic" networks), and fuel the decline of a community. Through this network analysis we have seen evidence that it is not just embeddedness or homophily alone but cohesion in its broader sense that is detrimental, and which can have implications on techno-economic outcomes. The Saliyars are, as they have been for a long time, over-embedded in a sense broader than the traditional definition of the term. While assessing their information and technological networks, agents must be cautious not only about being over-embedded but also about whether their embedded links are homophilous, too – for these have technological implications in the long run.

Notes

1 An earlier and much shorter version of this chapter appeared as A. Kamath and R. Cowan (2015), "Social Cohesion and Knowledge Diffusion: Understanding the Embeddedness: Homophily Association," *Socio-Economic Review*, 13(4): 723–746, https://doi.org/10.1093/ser/mwu024. Used with permission.
2 A master weaver is an entrepreneur of sorts who manages (often owns) a handloom textile manufacturing unit. A number of looms (ranging from just three or four to almost 100) are operated under one roof, employing labour and producing on a large scale.
3 A Geographical Indications (GI) Tag and its associated Geographical Indications of Goods (Registration & Protection) Act 1999, that India enacted as part of the WTO's TRIPS agreement, ensures that a product originating in and associated with a certain geographical region (such as "Bordeaux wine" to the Bordeaux region in France, or "Darjeeling Tea" to Darjeeling in India), is not produced elsewhere outside the region.
4 The princely state of Travancore occupied most of southern modern-day Kerala and a few regions in modern-day Tamil Nadu state. Travancore merged with the neighbouring princely state of Cochin and a former British province Malabar to form Kerala in 1949, the entire combination eventually merging with the Republic of India by 1956. Travancore had Trivandrum as its capital (also the capital of Kerala), 20 kilometres from which is Balaramapuram town.
5 Thanks are due to Neethi, again, for interpretation.
6 The ratio assesses degree of replication of ties in alliance groups and assesses in-group strength. A value greater than 1 shows that firms engage in more ties within the core group compared to outside it (Duysters and Lemmens, 2003: 60).
7 The correlation between two matrices is simply the correlation between elements of the adjacencies matrices. The element ij of an adjacency matrix takes the value 1 if i and j are linked in that network, 0 otherwise.

8 An interesting pattern in the significance tests is present. Differences in means between communities (in both networks) with regard to Non-Homophilous Embeddedness consistently were non-significant, while differences in means with regard to Non-Homophilous Embeddedness and Homophilous-Embeddedness consistently showed significance. The cohesion category in focus – Homophilous-Embeddedness – always showed strong significant differences between the Saliyars and others.

References

Arora, S. (2009) *Knowledge Flows and Social Capital: A Network Perspective on Rural Innovation*, Unpublished PhD thesis, UNU-MERIT and Universiteit Maastricht, The Netherlands.

Borgatti, S.P., Everett, M.G., and Freeman, L.C. (2011) *UCINET 6 for Windows: Software for Social Network Analysis*, Analytic Technologies, Harvard, MA.

Boschma, R. (2005) 'Proximity and Innovation: A Critical Assessment', *Regional Studies*, 39(1): 61–74.

Coleman, J.S. (1958) 'Relational Analysis: The Study of Social Organizations with Survey Methods', *Human Organization*, 17: 28–36.

Cowan, R. (2004) 'Network Models of Innovation and Knowledge Diffusion', MERIT-Infonomics Research Memorandum Series, MERIT, Universiteit Maastricht, The Netherlands.

Currarini, S., Jackson, M.O., and Pin, P. (2009) 'An Econometric Model of Friendship: Homophily, Minorities, and Segregation', *Econometrica*, 77(4): 1003–1045.

Duysters, G., and Lemmens, C. (2003) 'Alliance Group Formation: Enabling and Constraining Effects of Embeddedness and Social Capital in Strategic Technology Alliance Networks', *International Studies of Management and Organisation*, 33(2): 49–68.

Ghezzi, S., and Mingione, E. (2007) 'Embeddedness, Path Dependency and Social Institutions', *Current Sociology*, 55(1): 11–23.

GoK (2007) 'Karalkada: Elegance Manifested', *Kerala Calling*, February, 27(4): 20–21, Government of Kerala.

GoK (2018) *Kerala Economic Review*, Planning Board, Government of Kerala

Golub, B., and Jackson, M.O. (2009) 'How Homophily Affects Learning and Diffusion in Networks', Working Paper 35, Fondazione Eni Enrico Mattei, Milan, Italy.

Golub, B., and Jackson, M.O. (2012) 'How Homophily Affects the Speed of Learning and Best Response Dynamics', *Quarterly Journal of Economics*, 127(3): 1287–1338.

Granovetter, M. (1985) 'Economic Action and Social Structure: The Problem of Embeddedness', *American Journal of Sociology*, 91(3): 481–510.

Hanveev (2006) 'Diagnostic Study of Thiruvananthapuram Handloom Cluster', Report submitted to the Development Commissioner (Handlooms), Ministry of Textiles, Government of India, by P.S. Mani, Kerala State Handloom Development Corporation Ltd. (Hanveev), Kannur, Kerala, India.

Jackson, M.O. (2008) 'Average Distance, Diameter and Clustering in Social Networks with Homophily', in Papadimitriou and Zhang (eds.) *Internet and Network Economics*, Springer Verlag, Berlin and Heidelberg.

Kamath, A. (2015) *Industrial Innovation, Networks, and Economic Development*, Routledge, London and New York.

Kamath, A., and Cowan, R. (2015) 'Social Cohesion and Knowledge Diffusion: Understanding the Embeddedness: Homophily Association', *Socio-Economic Review*, 13(4): 723–746.

Krackhardt, D. (1987) 'QAP Partialling as a Test of Spuriousness', *Social Networks*, 9: 171–186.

Krippner, G.R. (2001) 'The Elusive Market: Embeddedness and the Paradigm of Economic Sociology', *Theory and Society*, 30(6): 775–810.

MoT (2010) *Annual Report 2009-2010*, Ministry of Textiles, Government of India

MoT (2018) *Annual Report 2018*, Ministry of Textiles, Government of India

Niranjana, S., and Vinayan, S. (2001) 'Report on Growth and Prospects of Handloom Industry', Study Commission by the Planning Commission, India.

Portes, A. (1995) 'Economic Sociology and the Sociology of Immigration: A Conceptual Overview', in Portes (ed.) *The Economic Sociology of Immigration: Essays on Networks, Ethnicity, and Entrepreneurship*, Russell Sage Foundation, New York.

Rajagopalan, V. (1986) *The Handloom Industry in North and South Kerala: A Study of Production and Marketing Structures*, unpublished MPhil thesis, Centre for Development Studies, Trivandrum, India

Raman, V. (2010) *The Warp and the Weft: Community and Gender Identities among Banaras Weavers*, Routledge, New Delhi and Abingdon, UK.

Rogers, E. (2003) 'Diffusion Networks', in Cross et al. (eds.) *Networks in the Knowledge Economy*, Oxford University Press, Oxford and New York

Soundarapandian, M. (2002) 'Growth and Prospects of Handloom Sector in India', Occasional Paper 22, National Bank for Agricultural and Rural Development (NABARD), Mumbai, India.

Uzzi, B. (1996) 'The Sources and Consequences of Embeddedness for the Economic Performance of Organizations: The Network Effect', *American Sociological Review*, 61(4): 674–698.

Uzzi, B. (1997) 'Social Structure and Competition in Interfirm Networks: The Paradox of Embeddedness', *Administrative Science Quarterly*, 42(1): 35–67.

Wasserman, S., and Faust, K. (1994) *Social Network Analysis: Methods and Applications*, Cambridge University Press, Cambridge, UK.

Woolcock, M., and Narayan, D. (2000) 'Social Capital: Implications for Development Theory, Research, and Policy', *The World Bank Research Observer*, 15(2): 225–249.

3

COMMUNITY SOCIAL CAPITAL AND INHERITED COHESIVE NETWORKS[1]

In Chapter 2, I explored the concept of homophilous-embeddedness, in the process of expanding the conceptual understanding of social embeddedness, to understand what undergirds techno-economic outcomes in the case of the Saliyars. In this chapter, I continue this study by examining how the homophilous-embeddedness in the extremely cohesive networks of the Saliyar caste-weaver cluster worked its way across generations, influencing a variety of economic and cultural factors, eventually driving the Saliyars of Balaramapuram into techno-economic decline. I demonstrate how homophilous-embeddedness among the Saliyars was actually a deep-seated attribute that helped carve their techno-economic outcome, and not simply an incidental characteristic of their modern-day business and technological information networks. I also present the mechanisms through which the absence of homophilous-embeddedness among the many other caste-heterogeneous clusters of weavers in this town stimulated their rise. At the conclusion of the chapter, I hope to show how affiliation to a rigid network and traits of homophilous-embeddedness can weaken even a seemingly prosperous group, regardless of industry performance and technological levels. At a broader level, this chapter also studies how complex social relations influence economic relations and technological progress, when these relations are relayed across generations in low-tech clusters. I begin by asking the seemingly simple question of why the Balaramapuram Saliyars cannot simply make their network links less cohesive, to tap new information sources and release themselves of their extreme caste cohesion.

Background

What is it that hindered an individual in the Saliyar community at Balaramapuram from amending his or her network contacts and including weavers and artisans of other castes in the town, especially when there is nearly negligible animosity among castes there? The answer lies in the Saliyars' perception of their social capital. The Saliyars treat their social capital almost as "ethnic" capital; many in this community strongly believing that weaving is "in their genes" and a matter of "community pride." We know from the literature that social obligations are deep-seated in the everyday economic functioning of communities. Networks may be the results of gradually solidified historical processes, iterated production rules, and communication protocols in interactions (Padgett and Powell, 2012). Inherited production links cannot be amended easily, and attempts to do so may be socially expensive, as it may involve fiddling with existing caste relations and tampering with investments made in the past by their community to maintain social ties and obligations with one another, specifically for economic purposes (Coleman, 1988; Borjas, 1992, 1995). "Cultural values," which often materialise in economic links, are often transferred across generations purely for their survival and preservation (Wintrobe, 1995; Dasgupta, 2005). Many Saliyars who were interviewed for this study reported that network links were ingrained into them while they grew up, with their parents and extended family familiarising with suppliers and consumers (essentially members of their own community) arriving at home every day. The baggage of loyalty and communal obligation was relayed generation after generation, "locking them in" from birth (Dasgupta, 2005). Information on links was directed by tradition, just as in a "network clan," where transmission of orders is based not on market signals such as price or on account of hierarchical commands but on account of traditions and informal regulations (Bianchi and Bellini, 1991). The Saliyars operated within such practices, too.

The expected problems with cohesive communities and ethnic enclaves – such as free riding associated with the public goods nature of social capital, or isolation in the larger population due to language or dialect – were bypassed in the case of the Saliyars. Free riding associated with the public goods nature of social capital was averted due to a strong presence of numerous closed networks within the community (note the many triangles in the Saliyars' networks displayed in the network diagrams in Chapter 2), and consequently the inescapable monitoring of each individual by the larger community. Also, both Malayalam and Tamil languages are freely spoken by the majority of the population in an interstate border town such as Balaramapuram, which is populated by many native (non-Saliyar) Tamil speakers. This surmounts any cultural isolation among the Saliyars within Balaramapuram.

Also, it is not the case that handloom was an unprofitable industry. The ongoing sustenance of the Payattuvila Cluster and many other such small clusters in Balaramapuram town and Trivandrum district show that handloom (though plagued with numerous other problems such as fluctuations, competition from powerloom, unorganised production, defunct cooperatives, etc.) has enjoyed a modest level of success, having also acquired a Geographical Indication tag for the Balaramapuram sari and for four other textile products, and catering to a strong product demand statewide and in upmarket showrooms across India. In fact, the literature has demonstrated that a unit's failings may not be on account of organisational issues or due to shortcomings in the industry but due to its position and affiliation to a cohesive and rigid network (Walker et al., 1997) – this seems to apply well to the Saliyars in the handloom industry, as we will disentangle in detail in this chapter.

The ultimate solution among the Saliyars to escape their inherited lock-in, extreme cohesive networks, and eventual decline was to abandon weaving and, in the long run, move away from the handloom industry altogether. But one must bear in mind throughout that this analysis of the Saliyars is not about why they moved from weaving to other professions; it is about the cause of their decline in the handloom industry at Balaramapuram, the root of which is found to be community cohesion and homophilous-embeddedness in their networks. This is not so much about why other professions appeared more promising or whether they were operating with obsolete technologies but how the Saliyars reached a dead end in weaving – their hereditary profession – due to rigidity and excessive cohesion in their networks, which was relayed across generations.

I organise and unpack these arguments by first presenting the proposition that community social capital has been central and congruent to technological progress in the handloom industry in India throughout the centuries. One consequence of this argument or claim is that the Saliyars, the exemplar in community bondage among weaving communities in Balaramapuram, should have actually progressed. This is backed by the evidence supporting the fact that because weaving, as a full time activity in handloom-engaged households in Kerala, was pursued with an intensity no lesser (and at times greater) than in India as a whole, the Saliyars were not specially disadvantaged either in the industry or in the region they are operating from – and should not out-migrate for good but rather exercise a flexibility to exit and reenter, as other communities have done in other states in India (see Mamidipudi et al., 2012).

> jumping off a home ship that is carrying too much load in bad weather, and swimming alongside on their own steam, till fair weather allows [weavers] to hop back on. There are casualties, of course, but the ship continues its journey, ferrying people from subsistence to sustainability.
> (Mamidipudi et al., 2012: 47)

While there are a multitude of cases in history demonstrating healthy relationships between community cohesion and technological progress among handloom weaver communities in India, in the case of the Saliyars the relationship, over time, unfortunately became antagonistic and unhealthy. The Saliyar out-migration from the industry in Balaramapuram is permanent, quite unlike the analogy given by Mamidipudi et al. (2012) above. To understand why the Saliyars are a counter example to the standard premise in the literature (which argues for harmony between community social capital and technological progress in handloom) in more ways than one, we are compelled to investigate into the centrality of community social capital among the Saliyars and inherited homophilous-embeddedness in their networks.

In this chapter, I first unpack the standard premise mentioned above. This is done by first perceiving handloom as a "socio-technology" and the weaver as a "socio-technologist," as well as by reviewing the evidence in the literature on the

49

congruent relationship between family/community centrality and technological progress in the handloom industry. This is then followed by a discussion based on data from NCAER (2010) and MoT (2012) that compel us to believe that because participation in weaving in Kerala has fared quite similarly to India in general, the Saliyars operated in an environment that is not significantly more disadvantaged than in the rest of India, and hence need not have permanently exited from the industry. The chapter then moves on to investigate what roles the centrality of community social capital and homophilous-embeddedness has played in the Saliyars' decline. In order to investigate this, we first survey the plethora of schemes and programmes that the central government, as well as state governments, provided to the handloom industry in order to stimulate its progress and growth. We then witness how the Saliyars did not participate in these, partly by excluding themselves from organisational innovations; how this led to their design information entering into a long-term phase of redundancy; and how issues of land subdivision plagued the sustenance of their functioning. As a parallel, we see how the absence of the possibility of community cohesion, and flexibility in networks, fuelled the rise of the other socially-heterogeneous clusters in Balaramapuram.

The centrality of community in technological progress in handloom

The literature on the Indian handloom weaving industry has shown that community cohesion and the adoption of innovations have historically played a *symbiotic* and not antagonistic role. To better understand this, we must first adopt the perception of handloom as a "socio-technology" and the weaver as a "socio-technologist" – a view expanded upon by Mamidipudi et al. (2012) in their analysis of weaver mobility in Andhra Pradesh state in India.

The handloom industry: a family/community-based socio-technological system

Mamidipudi et al. (2012) allege that the characterisation of handloom in India as static, traditional, and outdated, is unfair. Their study on handloom weavers in Andhra Pradesh challenges this notion, and urges us to appreciate that the handloom industry must be studied as an elastic and evolving *socio-technological system*. Technical functions in production are well-rooted within the structure and functioning of community, and the coexistence of the two is inevitable.

> Each weaving family . . . is linked to another five families through the auxiliary activities of dyeing, warping, sizing and winding. The weaving system is further linked to dyeing, credit and marketing through hybrid institutions that link rural and urban environments. This builds a com-

plex socio-technical and economic network that weaver households maintain and by which they are maintained.

(Mamidipudi et al., 2012: 50)

Weavers are aware, according to Mamidipudi et al. (2012), that their performance and technical expertise at almost every stage of production is correlated with their investments in social relations and in building social networks. The recognition that they speak a common technical and social vernacular, prompts weavers to mobilise knowledge within their social networks. By virtue of this, the weaver becomes a socio-technologist. This is demonstrated with the evidence from the literature on the history of technological progress in the handloom industry, which gives a sense of how innovation and information diffusion in handloom has always revolved around community, and has for the most part been positively stimulated by community social capital. I term this positive relationship the "standard premise" in the literature, and build it by reviewing the historical experience drawn from works primarily by Tirthankar Roy (1987, 1993, 1996, 1999, 2002) and Douglas Haynes (1996, 2001, 2012). It is against this standard premise that the experiences of the Saliyars of Balaramapuram are examined.

There is a common misconception on handloom in India, which has resulted in bestowing a pastoral notion on handloom textile production, picturing rural or small-town weavers as operating in a rustic (and somewhat static) setting. Haynes (2001, 2012) argues that handloom has, on the contrary, always been a dynamic industry, characterised by frequent innovation and weaver mobility; it has also always been a community-based industry, with entire communities engaging in it as their traditional profession. That is, handloom in India has been characterised by the household-based weaving family working not alone but embedded in community-based clusters that consider weaving (and generally handloom textile production from start to finish) as their community heritage and not simply a family's source of income, the community being the agency through which innovations have filtered into the industry.

Historically, the weaver or weaver family has rarely existed outside of community, a fact valid even to this day, where weaving in many regions in India is still community-based and caste-based at its core (Mamidipudi et al., 2012). Familial labour was always the basic unit of production in nearly all processes of handloom textile production (Haynes, 2012). And in terms of adoption of weaving innovation, the first adopters of innovations (such as electrification and modernisation of weaving, pre-loom, and post-loom processes) were communities drawn from hereditary "weaving castes" such as the migrant *Padmasalis* and the Muslim *Momins* or *Julahas* (Haynes, 2001, 2012; Roy, 2002). What makes this fact more interesting is that even ancillary activities were caste-based, distributed across different castes. Haynes (2012) provides a number of examples: dyeing castes regularly developing new knowledge of preparing fast dyes; other castes processed different kinds of coloured thread or cloth; gold-thread manufacturers of a particular caste also developed incremental innovations on their machines in the early 1900s; carpenters (again, of the carpenter castes) were making improved dobbies, English healds, and other weaving accessories; yarn preparation also saw incremental innovations.

The central place of caste and community in handloom has, according to Roy (2002), long been recognised in Indian policy, even during the British Raj. Roy explains that many innovations were intended to be introduced top-down by British administrators in the Indian subcontinent who were interested in population estimates of craftsmen in handloom regions – estimates which could be unearthed only through regular *caste* censuses (and not industry surveys, as would be the case in most other sectors).[2] These administrators had also discerned the fact that caste-based professions shared one (and probably the only) feature of the European guild, that of exclusive unity, which had connotations of collective information sharing and hence encouragement of absorption and diffusion of new information (Roy, 2002).

Let us look at Roy's development of this argument a little closer. The nuances of production in handloom were known only within certain castes, and the communities and clusters that were based on these castes. These provided (and still do provide, though in a reduced capacity) a social bond and distinct identity that influenced its members to channel profit to the common good by building community centres, temples, and so on. But very often, they also obliged members to share technical information and teach their progeny the profession that the caste was associated with, assist others in the community with production and technical problems, and at the same time restrict outsiders from all of this. Learning in handloom had a "strong apparent correlation with collective social identity" (Roy, 2002: 527). Cooperation, trust, assistance, and learning were, hence, through informal channels of communication in the community clusters, demarcated by social boundaries of caste and community.

In fact, the centrality of the family and caste in Indian weaving was always resilient, evident in English records dating to very early periods such as the late 1700s, which detailed rather meticulously how tedious it was in the beginning to penetrate long-existing community networks and caste hierarchies and enter into direct relationships with weavers (Arasaratnam, 1980). It took much longer than the British had expected to get direct access to (and therefore control of) the weavers. The sense of community was so strong that weavers were known to simply evacuate entire villages and migrate to other towns to set up production whenever their caste structure and position were under threat by the new industrial systems that the British had introduced.[3] Arasaratnam (1980) also provides a detailed account with case studies of a total embeddedness of production relations in caste relations, in weaving clusters in villages on the Coromandel Coast (southwestern coast of India), and how, at times, social heads of communities who had absolutely no role in pre-loom or post-loom activity, administered over the community's production activity simply because they were heads of the local weaver caste.

The emergence of caste associations and community clusters were the direct consequences of weavers' migrations from the villages in the southern part of the subcontinent to larger towns such as Sholapur and Bombay in present-day Maharashtra state, which, Roy (1999) explains, served as an important feature of the strategy of migrant weavers to "establish themselves economically and redefine themselves socially" (p. 72). Even within cities, caste groupings were so strong that their agglomerations assisted in promoting and maintaining their ritual life (Haynes, 2012). Recreation of community and regeneration of roots characterised these migrant weaver communities, who faced a need to collaborate and create a "common good" but at the same time compete (Roy, 1999).

Often, these migrations, and the final destinations of these migrant weavers, were assisted and directed by well-to-do patrons with political acquaintances. Weaver communities would have only welcomed this, according to Haynes and Roy (1999), as patronage by nobility and migrations of weaver communities were symbiotic: clothing being a means of defining social status, and association with the aristocracy bringing the weavers social and ritual privileges over and above what they had been endowed by the caste system.[4] It has be noted that through all the migrations, the organisational structure of these weaving communities changed only marginally, with caste identity always at the centre.

When things slowly began changing in the Indian subcontinent after the 1860s with the systematic introduction of organisational innovations in the handloom industry, such as workshops (the *karkhanas*) consisting of dozens of looms with paid labour and formal systems of production of handloom cloth and delivery for export markets (primarily Britain), the family/community-based economy still stood strong and resilient. Haynes (1996), who has studied this in detail, says that the reason for this was the initial fear among weaver communities of disruption of traditional production and delivery systems, and a fear that the new system might hinder surplus creation. Venkataraman (1935) explains that the introduction of the workshops also placed in front of the traditional family/community system an unfamiliar work environment that involved specific work hours, punctuality in arriving at work every morning, wages on a monthly basis for senior workers and on a piece-rate basis for weavers, and so on. This may have caused workshops to appear, at first, unattractive to caste-weavers (i.e., those weavers for whom it was a hereditary community profession) and attractive only to those who belonged to non-weaver castes. This was the case, Venkataraman documents, in the northern part of present-day Kerala state in the early decades of the twentieth century. In the Madras Presidency (comprising most of the southern peninsula) too, attempts by the British-run Industries Department to set up government-sponsored handloom factories were not successful on their introduction, as caste-weavers would not accept a workday governed by a clock (Haynes, 2012).

But the initial resilience was overcome, and the workshop form of organisation was eventually absorbed; not displacing the family/community system but instead growing alongside it (Haynes, 2001). According to Haynes, who has documented in detail the entry of workshops in the industry in Western India, the family slowly began incorporating the management of the workshop and marketing of produce into its existing division of labour. In regions such as central Tamil Nadu state (which was, like western India, a thriving textile centre in the subcontinent), caste-weavers were still dominant by virtue of comprising most of the workforce in the workshops that developed there. Caste and community monopolies in various artisan and other occupations in India, which underwent an eventual breakdown during the British Raj, did not seem to affect the handloom industry very much, as seen in the Madras Presidency where for over two-thirds of weavers, handloom textile production continued to be an entirely hereditary and community-centred activity (Venkataraman, 1935).[5]

Haynes (2012) describes how *karkhandars*, the chief operators of these *karkhanas* or workshops, were essentially people from weaving communities, who happened to be wealthier than the average household weaver and more enterprising in terms of diversifying their clientele. His meticulous analysis of the workshops as essentially family-based and community-based is presented as follows. As families with strong preexisting craft skills, *karkhandars* and their families were able to seamlessly adapt their products and·practices to demand shifts, by building on personal and community relationships to develop more reliable workforces. By patronising community causes and by creating cooperative societies, they established themselves strongly, and put in enormous efforts to prevent the skills of weaving from leaking out of their community. The labour for these workshops was, hence, sourced from existing weaving castes. In smaller workshops, the owners, their wives, and their children continued to weave and perform other tasks alongside hired labour. Employing "outside" labour often involved hiring entire families, employing the female workers within hired families in preliminary processes and the men for weaving. So, though the wage labour was supposedly from outside the family in the workshop system, it was actually sourced mostly from within the community through informal networks of kinship, friendship, and neighbourhood (Haynes, 2001).

All this evidence might prompt the argument that workshops were units that fostered an extreme sense of community cohesion. This is not entirely incorrect, but the reason for their success was simply that despite being very cohesive, they were not opposed to experimenting with new machines or designs from outside of the workshop and community cluster, or to expanding their client base and networks well outside of the community. In fact, the very fact that the workshop coexisted alongside the family/community system is what laid the path to the adoption of one of the most significant innovations in the textile industry in the Indian subcontinent – the fly-shuttle loom – in the late 1800s and early 1900s.

Innovations such as the fly-shuttle loom found favour, gradually, among weavers in regions such as the western part of the Indian subcontinent, as they did not appear to disrupt the division of labour in weavers' families. The *Padmasali* community weavers (also the *Momins*), for example, were said to be exemplars in introducing the fly-shuttle, even bringing in migrant labour from their home region of Telengana (present day northwestern Andhra Pradesh state) to operate on the new looms. It is of interest to note that this innovation was adopted by many well-off weaver households in the western Indian subcontinent much before the formal top-down introduction by the British (Haynes, 1996), only after which did it eventually move into the workshops on a much larger scale.

> We must be careful about attributing too much of the credit for change to the impetus of government. At the time Bombay began to introduce new kinds of loom, processes of ethnological transformation were already under way in many centres where the *karkhanas* predominated. We have already seen how . . . several workshop owners in Sholapur had adopted the fly-shuttle loom before the government had decided to disseminate it. In Surat, local weavers largely abandoned the traditional throw-shuttle loom for the Hattersley loom, . . . without any encouragement from the state.
>
> (Haynes, 2012: 215–216)

The workshops adopted these new technologies into their scheme of activities slowly and carefully, testing out their impact not only on fluctuating market conditions but also, importantly, on existing family and community relations. The progress from pit-loom to fly-shuttle, and in some cases even to powerloom, was by means of using this cautious and meticulous approach, attempting to maintain the long-existing division of labour based on family and community. The *karkhandars* who graduated to electric-powered powerlooms in the 1920s and 1930s were hereditary caste-weavers, once again relying on their own social groups. These *karkhandars* and their staff relied on trial and error to learn the new machines, and called upon relatives or neighbours who often helped them solve mechanical difficulties (Haynes, 2012: 251).

According to Haynes' (1996) assessment, the division of labour in 1940 (when the workshop form of production was strongly developing and operating almost entirely by fly-shuttle loom) was in fact not very different from that which existed in about 1900 (when these large process and organisational innovations were just being introduced). Also, despite the workshops being very successful in production and adoption of innovations, they never entirely displaced small family-based household units, even by the mid-twentieth century (Haynes, 2012). In fact, even in Independent India, by the 1960s, it was found by surveys and policy reports such as one by the Planning Commission (GoI, 1967) that the principal establishment in the handloom industry in India was still the weaver household and the principal workers of the industry were still weaver families. So handloom remained for the large part, in the late 1960s, still a hereditary and community-based industry.

> handloom weaving is a hereditary industry where the son learns from the father the techniques of weaving. . . . The handloom industry belongs to the traditional community of weavers. . . . Even after the advent of modern techniques and the growth of cooperative institutions the hereditary nature of the industry has hardly changed.
>
> (GoI, 1967: 17)

This was the case even by the 1980s; technological changes were found not to have fundamentally altered production organisation, with the household and family labour at the centre in most of the industry (Raman, 2010).

Hence, whether in migration or in the adoption of organisational and technical innovations, and whether in workshops or in households, family and community have always been the pillar around which handloom developed in most of India.

To reiterate the argument that sets the standard premise in this analysis, community social capital and technological progress have reinforced each other, very often favourably, and have shared a more or less symbiotic relationship in the handloom industry in India. If this is the case, it should therefore follow that for the Saliyars of Balaramapuram too, especially by virtue of being a migrant weaver community with official patronage and a multitude of other social and economic benefits, this harmony should have been long lasting. But this has not been the case. So can their sustained downfall be due to bad industry conditions in the state? Is handloom a sick industry in Kerala, and due to this, is weaving not the preferred activity for handloom-engaged households in the region? We answer this in the following section.

Participation in weaving in Kerala, compared to India on average

The handloom industry in India, as well as in Kerala state, is known for its uncertainties and fluctuations, as seen in production trends (Table 3.1).[6] But if we compare the situation in Kerala to the general situation across India, we see that many aspects of *participation in weaving* in Kerala fared quite similar to India in general. This prompts us to believe that the Saliyars operated in an environment that need not necessitate a permanent exit from the industry.

We shall see from this section that despite high fluctuations in the industry in Kerala, participation in weaving in Kerala was similar, or in some instances even better, than in India in general. I support this argument based on information

Table 3.1 Production of cloth in the handloom sector in India and Kerala state

Year	Production in India (million sq. metres)	Growth in production in India (%)	Production in Kerala (million sq. metres)
2002–2003	5980	–	70.75
2003–2004	5490	–8.19	56.82
2004–2005	5722	4.23	–
2005–2006	6108	6.75	62.38
2006–2007	6536	7.01	62.48
2007–2008	6947	6.29	70.88
2008–2009	6677	–3.89	20.20
2009–2010	6806	1.93	23.95
2010–2011	6949	2.10	–

Source: Based on Table 3.3 in MoT (2012) and GoK (various)

in the *Handloom Census of India 2009–2010* (NCAER, 2010), a comprehensive and broad-ranging report on various aspects of the handloom industry in India. This was the third such census to be produced in India, the second having been undertaken in 1995–1996 (hence the frequent reference to this year) and the first in 1987–1988.

Let us first look at handloom at an all-India level. At an all-India level, the majority of households associated with handloom cloth production were engaged at the weaving stage. This is a large majority of around 82% (numbering around 2.27 million households). Also, most individuals weaving in these households were not aged members of the family practicing an outmoded economic activity; in fact, 70% of the workforce was in the age group of 18–45. Though the population of weavers in India may have slightly declined, from 3.3 million in 1995–1996 to 2.9 million in 2009–2010, the proportion of *full-time* weavers among the total population of weavers actually increased significantly, from around 44.3% to around 63.5%. This goes along with a decrease in the number of idle looms among total looms in the country, from 10% in 1995–1996 to 4% in 2009–2010. Table 3.2 displays these and a few other indicators that show progressive figures. With these figures, we can judge that even if the handloom industry faced fluctuations over the period 1995–1996 to 2009–2010, weaving itself did not become a redundant activity to permanently move out of.

Let us now move to some closer aspects. The data is available at the state level for Kerala and at the all-India level but not at the district level for Kerala. With this limitation in mind, we move to Table 3.3, which shows that the proportion

Table 3.2 Comparison of selected indicators from the second and third handloom censuses

Indicator	Second census (1995–96)	Third census (2009–10)
Person-Days worked per Weaver	197	234
Share of Full-Time Weavers to Total Weavers	44%	64%
Share of Idle Handlooms	10%	4%
Share of Weaver Households reporting less than 1 metre of production per day	68%	46%

Source: Based on Table 10.15 in MoT (2012)

Table 3.3 Weaver or allied households as percentage of total handloom households (2009–2010)

	Weaver households (%)	Allied households (%)	Others (%)
Kerala	81.80	18.04	0.16
India	81.49	14.05	4.46

Source: Author's computations based on Table 3.1 by NCAER (2010)

of weaver households among total handloom-engaged households was nearly the same in Kerala as it is in India.[7]

And if we take a look at the workforce among households engaged in handloom, we see in Table 3.4 that the proportion of weavers among total handloom workers in households was around three-fourths, well past an absolute majority, in both Kerala and India in general. In fact, Kerala even enjoys a very slightly higher proportion of weavers in handloom households.

But it can be doubted as to whether these weavers, who seem to comprise the majority of handloom workers in Kerala as well as India, were engaged only on a part-time basis. If this is the case, we can be wary of the figures in Table 3.4 and judge that the industry was populated by individuals who wove as a peripheral activity, besides other economic activities that may be more rewarding.

But Table 3.5 refutes this, as we see that almost the entire population of weavers in Kerala worked full time in this activity (as do allied workers, and handloom workers in general). In fact, this proportion is much greater in Kerala than in India on average (where it is 63.5%). Also, there are more part-time allied workers than part-time weavers in Kerala, suggesting that weaving in Kerala enjoys a greater full-time participation than allied activities in handloom production.

Another indicator we can use to judge participation in handloom activity in Kerala is the number of workers in various categories of days worked per year.

Table 3.4 Proportion of weavers and allied workers to total workers in handloom in households (2009–2010)

	Proportion of weavers (%)	Proportion of qllied workers (%)
Kerala	76.97	23.03
India	75.61	24.39

Source: Author's computations based on Table 4.2 by NCAER (2010)

Table 3.5 Handloom workers by nature of engagement as percentage of total workers in each category (2009–2010)

Category of worker	Engagement	Kerala (%)	India (%)
Handloom workers	Full Time	97.37	64.26
	Part Time	2.63	35.74
Weavers	Full Time	99.02	63.49
	Part Time	0.98	36.51
Allied workers	Full Time	91.84	66.42
	Part Time	8.16	33.58

Source: Author's computations based on tables 4.9, 4.10, and 4.11 by NCAER (2010)

Here, in Table 3.6, we can see that this is the only indicator where Kerala per-formed a little below India on average, as the maximum proportion of handloom worker households (out of total – weaver and allied – households) feature in the category of 201–300 days worked per year, compared to the >300 category for an all-India level.

We can also see in Table 3.7 that the average number of person-days worked per year in Kerala by weavers was actually more than that of allied workers, though the situation is the reverse for India as a whole. More broadly, the average person-days worked per year by a handloom-engaged household was greater in Kerala than in India on average.

We now move to a critical indicator of participation and performance of weaving households among handloom-engaged households: the average earn-ing per annum. Table 3.8 shows, very clearly, that weaver households in Kerala reported greater average earnings per year than allied households in the state,

Table 3.6 Proportion of handloom worker households by number of days worked per year (2009–2010)

	<7	7–50	51–100	101–150	151–200	201–300	>300
Kerala	0%	1.08%	1.41%	3.14%	12.03%	61.97%	20.38%
India	0%	2.92%	15.49%	14.47%	16.30%	24.21%	26.55%

Source: Based on Table 4.15 by NCAER (2010)

Table 3.7 Total and average number of person-days worked per year (2009–2010)

	Average days per handloom-engaged household	Average days per weaver	Average days per allied worker
Kerala	296	246	214
India	264	183	217

Source: Based on Table 4.13 by NCAER (2010)

Table 3.8 Average earnings of weaver and allied households per annum (2009–2010)

		Weaver households (rupees per year)	Allied households (rupees per year)
Kerala	Rural	43,823	38,205
	Urban	31,242	29,571
	Total	**41,198**	**34,496**
India	Rural	38,260	29,693
	Urban	33,038	27,194
	Total	**37,704**	**29,300**

Source: Based on Table 6.7 by NCAER (2010)

and far greater than either weaver households or allied households at an all-India level. Contrary to what one might expect, weaver households in *rural* Kerala seemed to report the highest average earnings among all categories in Table 3.8.[8]

Hence, handloom – and particularly weaving – in Kerala has fared no worse than in the rest of India, and in some aspects even better.[9] Fluctuations in demand and other such problems plague the handloom industry as much as any other traditional industry in India, but weaving as a preferred profession in this industry has not taken a setback in Kerala. In Balaramapuram too, weaving as a profession has survived among socially-heterogeneous clusters of weavers who face the same industry conditions as the Saliyars. If pre-weaving and post-weaving processes, or even non-weaving alternatives, were more attractive than weaving, there should have been a mass migration of communities out of weaving. But this has not been the case. Even if there has been a general out-migration towards other professions, it may be in the manner that was expounded by Mamidipudi et al. (2012), where exit from and reentry into the profession characterise the migrations in and out of the industry. Weaver communities such as the Saliyars in Balaramapuram should not, ideally, have quit permanently but rather have exercised a flexibility to exit and reenter the profession, such as what traditional weaver communities in other states in India have been doing.

Mobility such as this, according to Mamidipudi et al. (2012), is the very basis of the maintenance of stability in handloom weaving and sustainability of the networks it operates within. So why did the Balaramapuram Saliyars, operating in a state whose participation in weaving is not worse off than the rest of India, choose to follow a one-way exit?

I argue that it matters only second whether the industry is performing well or not, as affiliation to a rigid network and traits of homophilous-embeddedness in the network can weaken even a seemingly prosperous community, even if operating in a modestly performing (or maybe even well-performing) low-tech industry cluster.

Understanding the Saliyars of Balaramapuram

The Saliyars are evidently a counter to the standard premise on the historically positive relationship between technological progress, industrial performance, and community. To understand why, I invoke the principal finding in the network study in the previous chapter – the presence of heavy homophilous-embeddedness in the Saliyars' networks, relative to the networks of other socially-heterogeneous groups (such as the communities in the Payattuvila cluster). The property of homophilous-embeddedness in a network delivers its outcomes in a convoluted manner, working its way by distributing its implications on a range of economic and cultural factors. It has implications on informal information sharing in these localities, on information diffusion, and ultimately on technological progress.

To compare with the Saliyars, I sketch how the other, socially-heterogeneous clusters in regions such as Payattuvila, who are currently enjoying a reasonable level of success, surged ahead over the decades primarily due to negligible community cohesion and homophilous-embeddedness in their networks.

But before drawing these paths, I first trace the events that transpired in the handloom industry in Balaramapuram, around the Saliyar cluster, in the 1960s and 1970s, from whence the Saliyars reported that their decline commenced. I describe a series of significant policy-prescribed developments from the 1960s onwards in the Indian handloom industry. It is after describing these policy efforts that I sketch the path through which the Saliyars' homophilous-embeddedness and community cohesion have worked their way through an assortment of mechanisms in the economic choices and functioning of the Saliyars over the last four decades, bringing the community down to their current deteriorated condition, and ensuring a long-term status to their condition.

State support for organisational innovation and for development and diffusion of innovative design

We must take a quick look at the various schemes and programmes developed by the state to serve what is probably the most important element in handloom and handloom technology – design. We must visit the state's efforts in in modernising the industry and planning efficient networks for innovation and diffusion of new information. Many of these were opted out of by the Saliyars, in order to maintain their rigid networks and community cohesion.

The broad production technology in weaving in Balaramapuram has remained essentially unchanged for around a century now, the antiquity of the technology being the basis of the very existence and consumer demand in this industry. However, within this broader technology, there have been numerous incremental innovations, especially in pre-loom activities such as spinning and winding/warping. The fly-shuttle loom was introduced in the Indian subcontinent in the early 1900s as an improvement over the pit loom, but both technologies operate side-by-side in this industry in Balaramapuram, each used for different products.

The information on production and new incremental technologies that circulates in these interpersonal networks revolves around the most central element in weaving – design. Success, according to Saliyar community members and weavers at Payattuvila who were interviewed, is said to come to those who have quick access to new information on the demands and trends in innovative designs. The individuals or groups who surge ahead are those who have access to vital nodes in the information networks that carry the information on innovative design and the method of producing these designs on the cloth. This was in fact recognised by the state, even in the 1950s, the first decade of policy planning in India after Independence. The government, at both central and state levels, felt the need to intervene in all three sectors of the handloom industry – cooperative societies, master weavers, and individual households – to promote design development and to universalise speedy access to new technologies around innovative designs. The Government of India sought to do this by establishing two Institutes of Hand-loom Technology (IHTs – one in Varanasi in north India, and another in Salem in south India) and several Weavers' Service Centres (WSCs – located all over the country), who were in turn advised to connect directly to the weavers and workshops in their respective region. The locations of the WSCs were very care-fully chosen in each state, ensuring proximity to the weaving hubs in the state. The government pursued the regular revision and reorganisation of syllabi at the IHTs which were at the apex of design development in the country, and which were to deliver the innovative designs to the WSCs through regular short-term training courses and exhibitions.

The WSCs were instructed to maintain close contact with exporters and pri-vately owned marketing organisations for information on modern fabric develop-ment, changing fashion demands, and other information. The IHTs and WSCs were to serve, in the language of our analysis, as state-led IIAs to assist in the effi-cient and ubiquitous diffusion of design information in their respective regions. The path that was charted for information on new design innovations was from the IHTs to WSCs, to proximate master weavers and cooperatives, and then to the individual households who were connected in some capacity to the master weavers and cooperatives. This was not without constant feedback between these actors and other significant private players in the industry.[10]

Besides this, the government also promoted modernisation and design devel-opment services for individual weavers who were outside the cooperative and master weaver fold, as well as for underperforming master weavers. A "High Pow-ered Team on the Problems of Handloom Industry" (whose report is MoC, 1974) had in this regard recommended the organisation of 25 units, each comprised of around 10,000 handloom weavers outside of the cooperative and master weaver fold in handloom hubs around the country, to receive training in new design, receive credit from nationalised banks, benefit from marketing of output, and strengthen linkages to WSCs.

In line with these propositions, by 1976, a Common Facility & Design Centre for weavers was set up in Kerala, in Balaramapuram. This had the explicit intent of promoting design innovations, providing training to weavers in design and technical advice in dyeing, printing, and other pre-loom and post-loom processes (GoK, 1976). This had its roots not only in the vast programmes for handloom development discussed earlier but also in the Government of Kerala's contribu-tion to the Twenty Point Programme announced by the Prime Minister of India in 1975. The state government had proposed two projects in Kerala (in the north in Kannur district, and in the south in Trivandrum district) for the intensive development of the handloom industry in the state, under the management of Hanveev (The Kerala State Handloom Development Corporation Ltd.). These projects were infused with funds as large as Rupees (Rs.) 1.85 million (in 1976), mostly with assistance from the Government of India. This involved the organ-isation of almost 100 workshop-type weaving units, the establishment of 100 collective weaving centres, and their linkage with the two WSC training centres in the state for design evolution and other technical issues (GoK, 1976). In the same decade, a large volume of funds (to the tune of Rs.11 million in 1976) in the form of cash credit was injected as working capital under the scheme of the Reserve Bank of India (the country's central bank), targeted not at house-hold weaving units but primarily at those who were under the cooperative or the workshop/work-shed form of organisation (GoK, 1978). For individual weavers, commercial banks were directed, under the supervision of Hanveev, to provide aid under differential interest rate schemes. These projects and the financial assis-tance that they brought along were continued beyond even the mid-1980s in Kerala (GoK, 1986).

In this manner, for around three decades – the 1950s, 1960s, and 1970s – there was intensive involvement of state support in this industry, concentrated in and around the handloom hubs in each state in India, including Balaramapuram.

Though a variety of such recommendations were provided by the central and state governments with regard to innovation and diffusion of design information, it was found in a study by the Planning Commission (GoI, 1967) that the fastest absorbers of new design information in the industry during the late 1950s and early 1960s were those who had also implemented the prescribed organisational innovations: namely the cooperatives and, more importantly, the *master weav-ers* and the *workshops*. It was revealed also by MoC (1974) – the report brought out by the "High Powered Study Team" – that though the *cooperative* mode of organisation was promoted by the government, the bulk of design development, the element that fuels the progress of the industry, came from the private sector, namely the master weavers who operate workshops and work-sheds and who were in close association with design development centres that were developed by the government during the 1950s and the 1960s. It was the master weaver, in other words, the one who adopted the *workshop* or *work-shed* mode of organisation on a large scale, who was said to have played a leadership role in design innovation. Brief attempts to discourage this mode of organisation from some quarters in the government (based on some accounts that there was rampant labour exploitation in these work-sheds and workshops) were put down consequent to surveys which revealed that:

> it would be a serious mistake if at the present stage of development we try to abolish this [master weaver] sector. . . . Till the cooperative sector is sufficiently developed and is able to give full service to its members and come up at least to the level which the master weavers have reached, it will be against the interests of weavers [for the State] to interfere with this sector.
>
> (MoC, 1974: 12)

In fact, even the earlier Planning Commission study (GoI, 1967) found through their analysis of a small sample of workshops in handloom-producing regions such as Andhra Pradesh, Tamil Nadu, and Maharashtra that it was the workshops, more than the individual households, that effectively adopted many of the innovations in the industry.

> all of the 11 workshops had adopted one or more types of improved implements. Among different improved implements varnished/wire healds were adopted by all the workshops; steel reeds and warping machines in 9 out of 11 workshops. The majority of workshops adopted dobbies/jacquards and take-up-motion attachments.
>
> (GoI, 1967: 32)

Independent households, constituting the bulk of the industry, did absorb some innovations. But they evidently lagged behind households that had embraced other innovative forms of organisation – such as workshop and master weaver arrangements. Independent households that excluded themselves from adapting to these organisational innovations also ended up keeping themselves away from the valuable training offered by state-sponsored agencies. The Planning Commission study provided some very interesting revelations regarding the self-exclusion of hereditary-weaving independent households who refused to participate in organisational innovations:

> out of 1097 sample weaver households, 1068 had no trained member. . . . This means that a very few namely 29 sample households were reported to have been trained under the training programme. . . . On the whole, weavers did not generally take interest in getting themselves trained in the improved methods of weaving. . . . A large majority of weaver households were not even aware of the existence of training programmes. . . . About one third of the households felt considered that the training was *not necessary . . . they felt that their members engaged in the weaving establishments were already trained because the occupation was hereditary, and as such they did not require any particular training in the industry.*
>
> (GoI, 1967: 39–40, emphasis mine)

To return very briefly to a historical instance, even in Western India in the early twentieth century, the community-based and caste-led workshops or *karkhanas* were the actors most enthusiastic in adopting some of the crucial technological and organisational innovations in the industry, while caste-weaver households were far more hesitant (Haynes, 2012). Innovative products were said to have stemmed from those *karkhanas* who were not opposed to the new machines and designs, and always on the lookout for new markets in the region and beyond. At the same time, Haynes (2012) continues, many weaving households decided not to participate in these sweeping changes; for instance, *Sali* and *Koshti* community weavers, who generally operated in households, did not adopt the fly shuttle as much as the *Padmasalis* did.

A school that had opened to spread new techniques among the Salis around the same time simply failed to attract new students. In part because of their difficulties in taking up the new methods, the Salis and Koshtis found it hard to compete and eventually left for other occupations.

(Haynes, 2012: 217)

These revelations demonstrate that those who were willing to absorb organisational innovations benefited from being at the forefront of design innovation and, after the 1950s, received enormous financial support from the state. But these findings also provide a hint as to what the attitudes were among some closed communities. Though the Planning Commission survey did not involve Balaramapuram, these results give us very interesting leads towards the analysis that follows.[11]

The Saliyars were in some sense better than the communities that were surveyed by GoI (1967), as they had adopted some smaller incremental innovations such as mechanisation of spinning. But, as detailed in the next section, where they faltered was in that they neither participated in absorbing and implementing organisational innovations (hence depriving themselves of financial incentives and schemes from the state in the 1960s and 1970s), nor had effectively tapped design innovations, which are possible to access through interpersonal networks. Both these exclusions had their roots in homophilous-embeddedness and community cohesion, as we shall see. The sections that follow are based on the illustration in Figure 3.1.

The decline of the Saliyars

According to a few Saliyar elders interviewed for this study, the first cause of the decline of the Saliyars can be attributed to the fact that their information on design was increasingly becoming redundant from the late 1970s and early 1980s onwards, which happens to coincide with the period when the state and many other bodies were infusing finance and many schemes and programmes into the handloom industry in Kerala.

Redundancy in design information

During interviews with Saliyar elders, it was revealed that the Saliyars used to pride themselves on the designs that they came up with and the innovative methods they developed to produce those designs on the final cloth – so much so, that Saliyar weavers strove to keep information on these a community secret.

Designs were shared willingly within the community but kept at close guard so as not to allow them to seep out until, of course, the final product went into the market. In this manner, though a mild and subtle competition existed among weavers within the community for innovative designs and innovative methods

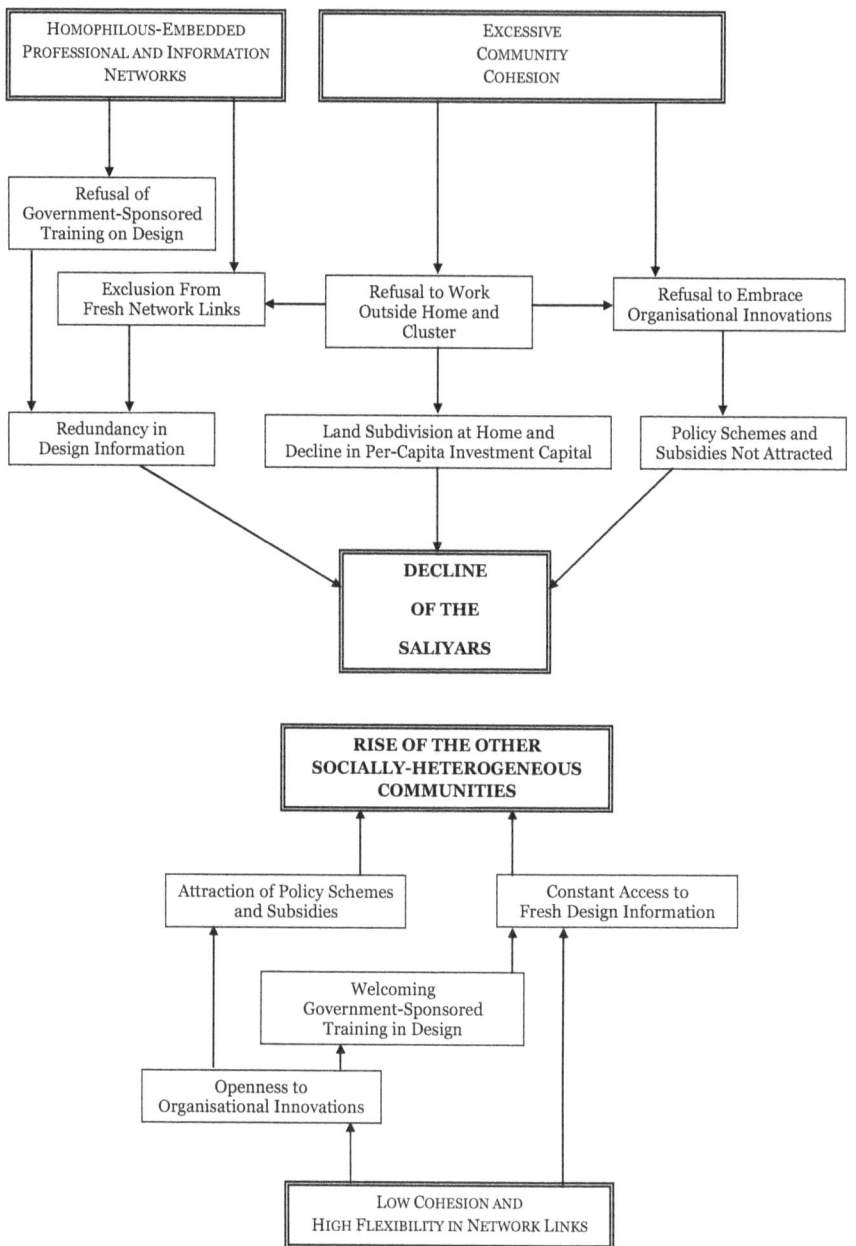

Figure 3.1 The decline of the Saliyars and the rise of the rest
Source: *Kamath and Cowan (2015)*

of generating those designs on the cloth, there was generally cooperation among Saliyar households to share information once a design had gained approval in the market. There was little input through information from the outside, as community cohesion was strong, and information networks were knotted mostly within the community.[12] This gives us pointers as to how and why redundancy in information began creeping into the Saliyars' information network.

Figure 3.1 illustrates that there are two main reasons why the Saliyars' information networks were plagued by redundant information: exclusion from fresh network links, and refusal to participate in government-sponsored training on design innovations.

Let us look at the first reason. The exclusion from fresh network links was due to the fact that the Saliyars' information links were inherited generation after generation, and each Saliyar household was locked in from birth to a network of suppliers, customers, and others who were the chief sources of information on new design. Being a network clan (Bianchi and Bellini, 1991), tradition dictated whom to ask and whom to talk to. Interviews with Saliyar elders revealed that the community would frown upon those who abandoned these traditional links, by distancing the deviant individuals during social functions and for production issues. This ensured a rigid network, which over a long term fed into an incapability to access fresh information.

The exclusion from fresh network links could have been averted if they had more efficiently utilised the opportunity that they had, in migrating to another village called Amaravila in the vicinity of Balaramapuram.

Amaravila is a tiny village eight kilometres from Balaramapuram, where a few Saliyar families established themselves from the 1970s onwards. This movement was not unidirectional but to-and-fro, with many families shuttling between Balaramapuram and Amaravila. By the late 1970s, Saliyar families in the Saliyar Cluster at Balaramapuram had begun to suffer from a problem common to agriculture in India – land subdivision. The sense of family, strong among weaver communities in India in general, was particularly deep seated in the Saliyars, so much so that Saliyar children would continue operating in the same household where they grew up and where their parents wove. With the area of the residence fixed, successive generations suffered from cramped households, and felt the need to move out of the Cluster. When a Saliyar family moved out of the Cluster, they wished to move only in the vicinity of Balaramapuram, and only to places where the community had possibilities to maintain a sense of identity and continue its religious and cultural practices. This was achievable where, for example, a temple with a favoured deity existed and where marriage relations were potentially possible with the existing inhabitants of the destination. Amaravila fitted these requirements very well, and so there was migration between this village and the Saliyar Cluster at Balaramapuram.

But what went wrong had roots in the same reason. Amaravila was not uninhabited, and had a small number of weavers from various communities. But the Saliyar families that moved to Amaravila were still attached to the home Cluster at Balaramapuram, sharing the same professional and information links. Hence, though there were a few weavers of different communities in Amaravila, the migrant Saliyars preferred to link with other Saliyars in their own home cluster at Balaramapuram. Links with these resident weavers of other communities could have begun the process of modifying the Saliyars' information network to include more out-of-community links, assisting them in slowly breaking out of their network rigidity. But the Saliyars missed this opportunity.

The move to Amaravila turned out to be a missed opportunity; it actually aggravated the inflexibility of the Saliyar network, by virtue of being associated with the same homophilous-embedded networks of the home Saliyar Cluster. Though a location change was undertaken with the intention of relieving themselves from land subdivision problems (which could have improved the structure and composition of the information network), the networks remained exactly the same – as did, therefore, the design information. In every sense, the Saliyars ended up operating in nothing but a new location attached to the same homophilous-embedded networks, rather than evolving fresh networks that could have arisen from the new location. In this manner, homophilous-embeddedness and a sense of community cohesion characterised links with Amaravila and fuelled the exclusion of the Saliyars from fresh network links for information on design innovations.

Another exclusion the Saliyars subjected themselves to was the training given to weavers in Balaramapuram (Kerala in general) by agencies such as the state, through training sessions organised by the nearest WSC as described earlier. This exclusion from training is interconnected with the refusal to embrace organisational innovations that attracted financial and technical support from the nodal agencies of the state, an issue discussed in the next section. The Saliyars willingly abstained from government-sponsored training workshops on design, detailed earlier in this chapter, as they had prided themselves on their capability to work as a community to come up with innovative designs and develop the expertise to weave those designs on the cloth. As one Saliyar elder put it, "we didn't need anyone from outside to tell us what to do."[13]

Very evidently, this standpoint maintained by the community, supported by social pressures not to participate, stemmed from an extreme sense of community cohesion. Hence, we see that homophilous-embeddedness in the Saliyars' networks and a sense of intense community cohesion was at the root of redundancy in information and thus impeded innovation on design, which is one reason that fed into the decline of the Saliyars.

Failure to adopt organisational innovations, and
to attract policy schemes and funds

Another reason for the community's decline, stemming again from homophilous-embeddedness and community cohesion, was its failure to attract policy schemes and funding assistance from the state. These schemes were an integral part of the state's assistance to the handloom industry which, as we have seen, continued for more than 30 years, beginning in the mid-1950s and carrying on beyond the mid-1980s. As we have seen, the state had, for a long period, very systematically drawn-out welfare schemes for training in design, funds for working capital and for purchase of new looms and other equipment, and substantial financial and technical support for the embrace of organisational innovations such as workshops. Workshops, usually attached to a cooperative, were also introduced to schemes for marketing. Failing to enrol with these ways of functioning and refusing to embrace certain key organisational innovations that characterised the handloom industry in Balaramapuram, was another factor that founded the Saliyars' decline.

Organisational innovations mainly involved the adoption of the work-shed (or workshop), which was attached to local cooperatives, and within which wage labour was employed on numerous looms. Supervision in these work-sheds was supposed to be under a master weaver who may or may not actually weave, and who played more of an administrative and advisory role, including the acquisition of new information on design and linking with the nodal agencies for design, either under the state or other private individuals. The master weaver, as studies have often found, was the chief agent in design innovation. It was revealed during interviews with the Saliyar elders that the work-shed and master weaver arrangements were not really brand new organisational innovations in the true sense of the term, as there were similar master weaver and workshop arrangements among the Saliyars (similar to the *karkhanas* within weaving communities in western India).[14] What was new in Balaramapuram was that organisational arrangements such as workshops, when introduced in this region, were prescribed as completely devoid of any community affiliation, an arrangement which the Saliyars had little agreement with.

There were, until the 1970s, according to Saliyar elders, almost 300 Saliyar master weavers in the community, employing a handful of workers at looms each of their homes. These weaving units with a cluster of looms were located within the Saliyars' homes, and employed labour from amongst the Saliyar community as well as from other communities. However, employees from the Saliyar community outnumbered those from outside by a large majority. Saliyar employees were, naturally, sourced from extended families or to maintain community relations, and worked inside the homes of the Saliyar master weavers. But the outsiders were allowed entry to, and operation from, only the backyard of the Saliyar home and not within the residence, where household members and other Saliyar employees worked. Moreover, employees from the other communities were employed not in weaving but in pre-loom and post-loom activities, which meant that they were expected to offer little in terms of bringing new information on the crucial element of design. Subdivision of land at home shrunk space to operate one's own family's production activities, let alone operate master weaver arrangements, and paying wage-labour became more difficult. These led to the slow disappearance of these Saliyar master weavers. Attempts to set up work-sheds outside of the

Cluster (where land was not scarce at the time) were rare, as most Saliyars reportedly did not want to leave the home Cluster.

Also, no Saliyar male was known to work for another caste's master weaver (for reasons of "caste pride," this justification cited consistently by those who were interviewed); and the Saliyar women who were employed at home for pre-loom tasks were in any case not permitted to work in handloom outside of the house. The Saliyars were very keen on sticking to their own organisational form – the household production unit with family/community division of labour – and their own cooperative societies demarcated along community lines, dense with homophilous-embeddedness.

Due to this environment in the Saliyar cluster, it was difficult for the Saliyars to break down their community cohesion in order to accept fully the work-shed form of organisation and production. Due to this non-acceptance, they were excluded from the links to the WSCs, and in turn the IHTs, the loans and funds from the commercial banks, and the many other policy schemes and programmes. The organisational innovations were simply not adopted, and the Saliyar master weavers receded.

It must be noted that the Saliyars had no misgivings about adopting innovations such as electric spinning and winding machines, new forms and variants of dyes and yarn, and other such small incremental innovations in pre-loom and post-loom processes. This is because these innovations were not at odds with community structure and functioning, and did not expect movement out of the home and the cluster. But merely the acceptance of these innovations did not ensure any progress, as organisational innovations were also crucial for survival and sustenance. This major innovation was the one that the Saliyars had backed out of, due to community cohesion. Hence, once again, community cohesion stands out as being at the root of the refusal by Saliyars to adopt organisational innovations (as enthusiastically as the other socially-heterogeneous clusters accepted them), leading to a failure in attracting policy schemes and funds.

Land issues at home and nearby, and decline of per capita investment capital

The Saliyars were endowed with large amounts of land when they were invited to Balaramapuram. This included not only their set of streets and their residences but also large tracts of land spread across a couple of acres around their cluster. This extra land was for a long time a principal source of finance for investment into the handloom business, and a source of financial security for the family. However, two issues arose as the decades passed: first, the increasing difficulty in employing and financing wage labour to maintain the economic activities operating in the extra tracts of land; and second, the division and sale of sections of land for marriage-related and dowry-related matters. Both of these were equally severe factors in depleting the stock of land that the families in this community owned. Other venues of sourcing investment included internal contributions from within the community, such as borrowing and lending from relatives and other acquaintances. Banks and other financial institutions were seldom considered a source of investment capital, despite the fact that agriculture and traditional industry such as handloom were targeted with massive financial support by nationalised banks. This meant that from the 1970s onwards there was an eventual diminution of sources for investment into the weaving business, adding to the difficulty in maintaining wage labour at home for weaving, and the eventual closure of Saliyar master weaver units by the early 1980s. These extra tracts of land, if still available to the Saliyars, may have even allowed them to continue, and set up new master weaver units that might have attracted funds and state-led schemes. But by the time the schemes were developed in Kerala in the late 1960s and mid-1970s, the Saliyars had lost most of their extra tracts of land.

Yet another land-related issue, briefly visited earlier, was the subdivision and partitioning of the house to allow successive generations and their families to weave at home. A decrease in per capita land at home meant that per capita production also fell, followed in turn by a fall in per capita investment into weaving, pre-loom, and post-loom activities. Naturally, expansions in weaving activity stopped and then began declining, handloom production among the Saliyars slowly beginning to incline towards pre-loom activities that required little investment and space to expand, compared to weaving. An extra loom, for instance, took up most of the floor area in one of the large rooms in an already crowded house, whereas an extra spinning machine took up very little space, as all that was needed was a little area on the side of a wall in a relatively smaller room. A yarn shop was easier to expand and the yarn business easier to invest in, given that extra bales of yarn took up very little space and the nature of the commodity's sale was fast moving (requiring less space requirements for inventory), compared to a master weaver arrangement that took up many times the area of an entire yarn shop. An increase in intensity in spinning or other pre-loom activities could have improved the Saliyars' condition, but an expansion in even these activities that required very little space had its own limits in an already small, and increasingly crammed, residence. Subsequent generations would set up a loom or spinning wheel in another quarter of their respective homes, this practice naturally reaching a limit within two or three generations. Given that this community was brought to Balaramapuram in the 1890s, one can picture that there would have been at least two or even three generations in the house by the 1970s actively pursuing weaving and other activities simultaneously, in the same small space of the home. Visits to Saliyar houses even today show spinning and plying activities in some Saliyar families literally jostling for space with day-to-day living arrangements.

All these again point at one root cause – the community cohesion that did not encourage (even if it did not strictly disallow) movement and operation out of the home and outside of the Saliyar cluster. Movement out of the cluster only meant towards Amaravila, the problems associated with which we have already covered in the previous discussion.

Summing up

These three factors – (a) redundancy in information on innovative design, (b) failure to attract state-sponsored schemes and funds by not embracing organisational innovations, and (c) diminishing land and investment capital – all appear to have their roots in two very characteristic traits or attributes of the Saliyars: homophilous-embedded networks and excessive community cohesion. As we have seen, they are highly interconnected and often overlap. The Saliyars began stagnating slowly by the late 1970s, and a crisis began building up in the Saliyar cluster from the early 1980s onwards, intensifying by the late 1980s and through the 1990s, and continuing even to the present day. This has its origins in the unwarranted degree of community centrality displayed by the Saliyars, manifested and fed-back by homophilous-embeddedness in their business and, more importantly, in their information networks. The Saliyars had recognised that their progeny had a choice between, on the one hand, inheriting the same networks, sticking back in the home cluster with a household form of production organisation and family/community division of labour, avoiding tampering with caste relations; and on the other hand, to leave weaving and handloom altogether. They seem to have chosen the latter.

The rise of the other socially-heterogeneous communities

The depletion of resources for further investment and the redundancy in information on design brought about the stagnation of the Saliyars, and the cooperative societies they were associated with. These Saliyar cooperatives, drawn on community lines, became slow and laggard in operation, many going defunct and existing only on paper by the early 1980s. This is seen clearly in DoH (1986), one of the works that was commissioned by the state government to assess the performance of the schemes listed earlier in this chapter, and the performance of the industry as such.[15] In this report it is noted that by 1984, the chief cooperative society of the Saliyars, the "Anchuwarnatheruvu HWCS Ltd.,"[16] was listed as "dormant" (DoH, 1986: 7). What is interesting is that this cooperative society was the *only* one that had a status listed as "dormant," while the multitude of other cooperative societies in Balaramapuram – comprised of socially heterogeneous groups of weavers – were listed as "working."

This dormancy of the Saliyar cooperatives led to the next step – the takeover of these Saliyar cooperatives by the state and welcoming associations with weavers who belonged to all communities, not just the Saliyars. The remodelling of these societies was not simply by means of dissociation with the Saliyar community and entry of the other communities but by the systematic mediation of the government, playing an active role in reorganising the management of the societies, reworking their functioning and, most important of all, rewiring the networks that they operated on.

This brought about a "secularising" of these formerly Saliyar-only cooperative societies, with all communities free to enter and participate. Of course, there were weaver households already in the areas around the Saliyar Cluster long before the Saliyars came in, but these were (according to the Saliyars) only small-time weavers and only a sprinkling in number, like any other village or town in India which had a small number of resident weavers. Also, there were a few small cooperatives in Balaramapuram besides the Saliyar cooperatives, but according to the Saliyar elders interviewed, it was mainly after the state's remodelling and nurturing of the cooperative societies in a big way that the other socially-heterogeneous clusters actually progressed from being modestly successful to thriving in the business. These socially-heterogeneous clusters of weavers gained access to information from many external sources, and importantly, from the state in its regular training sessions on design innovations. They were subject to the WSC training sessions, links with exporters and other private players, funds and loan arrangements through the handloom development project in Trivandrum in the mid- and late 1970s, and many other progressive schemes. However, taking over the societies was just the initial step, and the links with these state schemes were not the principal basis for the sustained rise of the other socially-heterogeneous clusters in weaving. The cooperatives that we refer to here also ended up in many quagmires over the decades, but the socially-heterogeneous clusters continued to rise and sustain themselves in the business.[17] Though these linkages and attraction of schemes and programmes boosted their businesses, their rise was due to two principal factors: (a) a convivial attitude towards organisational innovations (such as the work-sheds and master weaver form of organisation), and (b) the flexibility in information network links. Both these factors had roots in the fact that these communities operated in very socially-heterogeneous environments that allowed little possibility for excessive cohesion.

The other socially-heterogeneous clusters in Balaramapuram, whether as a part of the newly taken-over cooperatives or not, had welcomed the organisational innovations that were supported by the state through funds and training programmes.

This was one crucial difference between the Saliyar cluster and the Payattuvila cluster. It was noted even during fieldwork that while the former was characterised by home-based family/community-labour units, the latter was characterised by work-sheds (at home or around the residential area), which may have been owned by a family but which employed wage labour sourced from the town regardless of caste, paid on a daily-wage basis, and attached to the local cooperative. The latter organisational form had drawn support from the state for the purchase of looms for work-sheds, incentives in terms of tax benefits, welfare benefits for workers (those handloom workers attached to cooperatives), infusion of large amounts of working capital through nationalised banks, and many other such schemes. The other socially-heterogeneous clusters had far better access to more flourishing domestic and other markets across India, thanks to associations with state-administered bodies such as the WSCs, and began supplying to large upmarket showrooms in Kerala and India in a capacity far greater than the Saliyars. With these sources and the regular training sessions in design organised by the WSCs, the other socially-heterogeneous clusters in Balaramapuram were said to have surged ahead.

These communities, with no restrictions on network links and information sources, enjoyed a flexibility that the Saliyars had eschewed, out of community cohesion and rigid inherited networks. The socially-heterogeneous clusters had constant access to fresh information on designs, and were always up to date on the latest trends in the market. They could associate and dissociate with agents in their networks as they wished, as they had no community or family obligations to bear. They had no restrictions keeping them from moving out of their homes and expanding whenever necessary, and no traditions directing them on how to operate their business. To reiterate, even when the handloom cooperatives in Kerala began languishing, clusters such as the one in Payattuvila enjoyed a prolonged participation in the weaving business thanks to the welcoming attitude towards innovations – especially organisational ones – and network flexibility.

Even in Western India, a century ago, the family-run workshops or *karkhanas* were welcoming towards innovations and new designs. Reliability was an essential element in the workshops, for which community and family cohesion were depended upon. Innovation, on the other hand, required new information, and therefore periodically "refreshed" network links. The workshop clusters of Western India during the time appear to have taken advantage of both worlds – community cohesion in production relations inside the workshop, and external connections outside of the community with regard to contacting design innovators, looking to the state for financial incentives (when offered), new trade links and customers, and so on. Even division of family labour in these workshops was entrepreneurial. Members of the owner-family were involved in operations such as searching for new markets and customers, networking outside of the town, and furthering similar non-homophilous–embedded links. Embedded or homophilous links inside the workshop may have been safe, but the workshop-owning families knew that homophilous-embedded links outside of the workshop were dangerous. This was the crucial difference between the family-owned and community-based workshops in Western India and Payattuvila on the one hand, and the Saliyar Cluster on the other hand.

The machines and loom processes were never different (they still are not) between the Saliyars and a community such as Payattuvila. Both use the same kind of looms (after all, the employment of this weaving technology is the very basis of demand in this industry), the same spinning machines, the same dyes, the same beaming techniques, and so on. The big difference is in organisation of production activity and, importantly, network flexibility. Also, a point to bear in mind is that the home retains centrality, even in Payattuvila where master-weaver arrangements thrive. Master weavers in Payattuvila manage their work-sheds or workshops *in the vicinity of their homes* with considerable family involvement. The difference lies in the fact that they have not been opposed to involving non-family employed labour in the workshops, have been welcoming towards organisational innovations and, most importantly, have been operating on less homophilous-embedded information networks.

A key aspect therefore is whether the *networks* were flexible or not, and whether choosing expansion outside of the house and community was accepted by members of the community. A socially-heterogeneous environment, with no community-related baggage, was (and still is) beneficial to maintaining the flexibility of the networks, and is also open to the possibility of migration outside of the house and the community. The possibility of a dangerous degree of community cohesion, and of homophilous-embedded production and information networks, is very low in such environments, and was at the root of the rise of the socially-heterogeneous clusters in handloom weaving in Balaramapuram.

Conclusion

The Saliyars found themselves trapped. Though calamities in their Cluster have come and gone in the past, the dilemmas that have sustained and even exacerbated over the last 30 years are worse than in the decades prior to their decline. Handloom, and traditional industry in general, has always been infamous for the uncertainties and fluctuations that have often put its workers at grave risk. In fact, historically too, in times of crisis such as famine, it has been noted by the literature that weavers were the first to starve (Roy, 1993, 2002), and hence the first to migrate. Those communities that migrated were found to display an excessive, sometimes unwarranted, amount of community cohesion and centrality to family. But this community and family spirit is also known to have assisted and given its own shape to the technology trajectory of weaving in the handloom hotbeds of India. Inherited networks and community social capital overpowered the risks of adoption of new technologies and practices, and invigorated information flows. Community social capital and technological progress went hand-in-hand for the most part in the history of Indian handloom, as the former was used judiciously and not across all respects of production and information seeking. But while for the most part community cohesion has been, historically, congruent to technological progress and knowledge diffusion among community-based weaving clusters and groups in India, in the case of the Saliyars there has been a disharmony. This does not require rethinking of whether or not community social capital and technological progress share a healthy relation. They still do share a positive relationship, though only to a limit, after which the detriments of community social capital set in and rigidities associated with inherited networks set in, and hinder information diffusion and technological progress. But whether for better or worse, technological outcomes in handloom are pivoted within the sociological setting. The manner in which these settings operate is often complicated, and its analysis sometimes requires – as has been undertaken in Chapter 2 – reframing.

Notes

1 An earlier and much shorter version of this chapter appeared as A. Kamath and R. Cowan (2015), "Social Cohesion and Knowledge Diffusion: Understanding the Embeddedness: Homophily Association," *Socio-Economic Review*, 13(4): 723–746, https://doi.org/10.1093/ser/mwu024. Used with permission.

2 One could argue that the justification for caste censuses (as opposed to general population censuses or industry surveys) may be weak, on the grounds that weavers may generally belong to particular castes but that *all* members of that caste may not weave. There is some truth in this argument, but one must bear in mind that there was a slim chance of finding weavers in other castes, especially during the period in history involved here. Hence, a pragmatic way to capture demographic information of weavers was to conduct surveys in and around weaver *castes*, even if all caste members were not weavers. I provide evidence later in this chapter (from Venkataraman, 1935) that even until the middle of the twentieth century, and despite the slow breakdown of caste and community monopolies in various occupations during the British Raj and after Independence, the large majority of weavers and handloom textile producers continued to hail from hereditary weaving communities.

3 Though Arasaratnam (1980) does not provide details of which communities exactly did this, he puts forward a very interesting argument that the weaver responses of the 1770s and 1780s in South India around the Carnatic region (most of modern day Karnataka, Andhra Pradesh, and Tamil Nadu states) were the first popular reactions against British rule in India.

4 This is very similar to the Saliyars of Balaramapuram, though they migrated at a much later period and not out of circumstance but out of invitation of the Maharaja of Travancore, under whose patronage they worked. The Maharaja on the one hand invited them to ensure his supply of Saliyar-woven high-quality clothing, and the Saliyars on the other hand, with his patronage, lived a lifestyle and observed community practices that were much higher than what the caste system had traditionally accorded them.

5 Venkataraman finds that the *Kaikolars, Devangas, Salés* (not related to the Saliyars at Balaramapuram) and *Sourashtras* were still the dominant weaving communities in the Madras Presidency, well into the twentieth century, even after the large-scale producing workshops began seeping into the handloom industry.

6 Production in Kerala appears to have faced a sudden and substantial drop after 2007–2008. This might prompt the allegation that industry conditions were indeed worse in Kerala than in the rest of India, which may actually be at the root of the permanent exit of the Saliyars. But this argument cannot hold, as the Saliyar participation in weaving had already declined by then, and the community had largely exited handloom. If handloom had faced a sudden bad patch in Kerala during this period, or even in the past, the issue remains as to why it was mainly the Saliyars who left en masse, while others stayed and continued weaving. If there was a major decline in the industry, the Saliyars and other communities would have left in more or less equal proportions, other things remaining equal. But this has not been the case.

7 A *weaver household unit* is defined by NCAER (2010) as "one that has any member of the household who operated a loom even for one day in the last one year (preceding the survey date), either within the premises of the house (classifying the household as a 'with loom household') or outside the household premises (classifying the household as 'without loom household')." On the other hand, an *allied worker household unit* is defined by this census as "one that has any member of the household who has undertaken pre-loom (dying of yarn, warping/ winding, weft winding, sizing, testing, etc.) and/or post-loom activities (dying of fabric/calendaring/printing of fabric, made ups, etc.), even for one day in the last one year (preceding the survey date), either within the premises of the house or outside the household premises. These households did not

have any members engaged in weaving activity within or outside the premises, nor did they have a loom within their premises." NCAER (2010: 6).

8 It must be noted that these figures, if calculated at a monthly level, appear quite low in absolute terms especially when considering rising living costs in both rural and urban India. However, the objective here is to demonstrate that even if weaving is not a lucrative option in terms of money earning, it is still better off for the most part in Kerala than in India on average.

9 Though Kerala in some indicators shows a greater participation and intensity in weaving compared to an all-India level, it is by no means the primary handloom weaving state in India. Other states such as Andhra Pradesh, Maharashtra, and Tamil Nadu have far larger weaving industries compared to Kerala, in terms of output, export, etc. Evidence supporting this is replete in NCAER (2010).

10 There are today five IHTs (rechristened IIHTs – Indian Institutes of Handloom and Textile Technology) in Varanasi, Salem, Guwahati, Jodhpur, and Bargarh, as well as 25 WSCs in almost all the states. In addition to the IIHTs managed by the Central Government, there are, in addition, four handloom design and technology institutes managed by the state governments, in central and south India, including in Kannur in Kerala. There is also a National Centre for Textile Design (NCTD) at New Delhi.

11 In Balaramapuram, design innovations played a more significant role than technical innovations, unlike in the towns surveyed by the studies such as GoI (1967).

12 This recalls Burt (1992), who explained that redundancy of information leading to obsolescence is indicated by cohesion and equivalence, which manifest in network structure. Cohesive contacts, being strongly connected to one another, provide the same information repeatedly. Equivalent contacts, connecting an agent to the same third party, also direct the agent to receiving redundant information.

13 This is reminiscent of what was found by GoI (1967) in their survey, that many hereditary caste-weaver households did not find the training "necessary," by virtue of handloom textile production being their hereditary profession.

14 The *karkhanas* of Western India arrived more than a century earlier than in Balaramapuram, for a simple reason. While Balaramapuram was at the time a very small village with a sprinkling of weaver households catering to a local and domestic market, towns in the western part of the Indian subcontinent were at the heartland of textile production in the subcontinent, producing for export to Britain and other colonies.

15 DoH (1986) was an effort by the state government to list out in the form of a directory, the primary handloom cooperative societies in Kerala state and their status at the time.

16 "Anchuwarnatheruvu" is literally "lane of five castes." The Saliyars were invited to Balaramapuram in the 1890s by the then Maharaja of Travancore, along with a few families from four other Tamil-speaking communities. The main street on which the Saliyars – the most populous and prominent among these communities – were located was known as the "lane of five castes." HWCS is Handloom Workers Cooperative Society.

17 Despite cooperatives generally stagnating, it is the adoption of the workshop and work-shed form of organisation, operating under master weavers and having flexible and dynamic information network links (i.e., a welcoming attitude to knowledge from the outside) that helped sustain these socially-heterogeneous clusters, such as in Payattuvila, even until today.

References

Arasaratnam, S. (1980) 'Weavers, Merchants and Company: The Handloom Industry in Southeastern India 1750–1790', *The Indian Economic and Social History Review*, 17(3): 257–281.

Bianchi, P., and Bellini, N. (1991) 'Public Policies for Local Networks of Innovators', *Research Policy*, 20(5): 487–497.

Borjas, G.J. (1992) 'Ethnic Capital and Intergenerational Mobility', *The Quarterly Journal of Economics*, 107(1): 123–150.

Borjas, G.J. (1995) 'Ethnicity, Neighborhoods, and Human-Capital Externalities', *The American Economic Review*, 85(3): 365–390.

Burt, R.S. (1992) *Structural Holes*, Harvard University Press, Cambridge, MA.

Coleman, J.S. (1988) 'Social Capital in the Creation of Human Capital', *American Journal of Sociology*, 94: S95–S120.

Dasgupta, P. (2005) 'Economics of Social Capital', *The Economic Record*, 81(255): S2–S21.

DoH (1986) *Directory of Primary Handloom Cooperative Societies in Kerala*, Directorate of Handloom, Government of Kerala.

GoI (1967) *Study of Handloom Development Programme*, Programme Evaluation Organisation, Planning Commission, Government of India.

GoK (1976) *Economic Review Kerala 1975*, Planning Board, Government of Kerala.

GoK (1978) *Economic Review 1977*, Planning Board, Government of Kerala.

GoK (1986) *Economic Review 1985*, Planning Board, Government of Kerala.

Haynes, D.E. (1996) 'The Logic of the Artisan Firm in a Capitalist Economy: Handloom Weavers and Technological Change in Western India, 1880–1947', in Stein and Subrahmanyam (eds.) *Institutions and Economic Change in South Asia*, Oxford University Press, New Delhi.

Haynes, D.E. (2001) 'Artisan Cloth-Producers and the Emergence of Powerloom Manufacture in Western India 1920–1950', *Past & Present*, 172: 170–198.

Haynes, D.E. (2012) *Small Town Capitalism in Western India*, Cambridge University Press, Cambridge, UK.

Haynes, D.E., and Roy, T. (1999) 'Conceiving Mobility: Weavers' Migrations in Pre-Colonial and Colonial India', *The Indian Economic and Social History Review*, 36(1): 35–67.

Kamath, A., and Cowan, R. (2015) 'Social Cohesion and Knowledge Diffusion: Understanding the Embeddedness: Homophily Association', *Socio-Economic Review*, 13(4): 723–746.

Mamidipudi, A., Syamasundari, B., and Bijker, W. (2012) 'Mobilising Discourses: Handloom as a Sustainable Socio-Technology', *Economic and Political Weekly*, 47(25): 41–51.

MoC (1974) Report of the High Powered Study Team on the Problems of Handloom Industry, Ministry of Commerce, Government of India

MoT (2012) *Annual Report 2011–2012*, Ministry of Textiles, Government of India.

NCAER (2010) *Handloom Census of India 2009–2010*, National Council for Applied Economic Research, Government of India, New Delhi.

Padgett, J.F., and Powell, W.W. (2012) 'The Problem of Emergence', in Padgett and Powell (eds.) *The Emergence of Organizations and Markets*, Princeton University Press, Princeton and Oxford.

Raman, V. (2010) *The Warp and the Weft: Community and Gender Identities among Banaras Weavers*, Routledge, New Delhi and Abingdon, UK.

Roy, T. (1987) 'Relations of Production in Handloom Weaving in the Mid-Thirties', Working Paper 223, Centre for Development Studies, Trivandrum, India.

Roy, T. (1993) *Artisans and Industrialization: Indian Weaving in the Twentieth Century*, Oxford University Press, New Delhi, New York, and Oxford.

Roy, T. (1996) 'Introduction', in Roy (ed.) *Cloth and Commerce: Textiles in Colonial India*, Sage Publications, New Delhi, Thousand Oaks, and London.

Roy, T. (1999) *Traditional Industry in the Economy of Colonial India*, Cambridge University Press, Cambridge, UK.

Roy, T. (2002) 'Acceptance of Innovations in Early Twentieth Century Indian Weaving', *The Economic History Review*, 55(3): 507–532.

Venkataraman, K.S. (1935) *The Hand-Loom Industry in South India*, Supplement to the Madras University Journal, Madras, India.

Walker, G., Kogut, B., and Shan, W. (1997) 'Social Capital, Structural Holes and the Formation of the Industry Network', *Organization Science*, 8(2): 109–125.

Wintrobe, R. (1995) 'Some Economics of Ethnic Capital Formation and Conflict', in Breton, Galeotti, Salmon, and Wintrobe (eds.) *Nationalism and Rationality*, Cambridge University Press, Cambridge, UK.

4

SUBALTERN CASTES AND THE PROMISE OF ICTs[1]

While the last two chapters analyse the sociological contexts of techno-economic outcomes in handloom, taking the case of a cohesive caste weaver cluster in a small-town traditional industry in Kerala, we proceed towards the study of disadvantaged castes in a peri-urban setting in the relationship that these castes have with new digital communication technologies. From a network-historical disquisition we relocate to a more contemporary exploration of the complicated relationship that subaltern castes share with technologies such as mobile phones, reorienting our conceptual foundations from network analysis and homophily to the social construction of technological usage and the politics of socio-technological outcomes. From reading the policy efforts in traditional industry as facilitative in their intention, in this study we now reinterpret policy efforts in digital technologies as progressive in intention but naive in substance. From investigating into the decline of a once dominant and successful caste group, we move to disentangling the factors for a socio-historically distressed caste strata to be dodged by digital technologies, which are effortlessly accessed and provide an opening to supposedly "caste-neutral" digital spaces. With this study we traverse another realm in the task of improving our acquaintance with the social context of technological experiences in contemporary India.

Introduction

This study presents a fresh perspective on the complicated liaison between new technologies and disadvantaged castes, in a peri-urban setting. It aims to deepen the understanding of experiences of digital communication technologies, in the impact that these technologies have on the lives and livelihood of socio-historically deprived and excluded castes in peri-urban south Bangalore. Primary oral accounts from individuals hailing from Dalit[2] castes in the region assist in uncovering their engagement with new digital technologies such as mobile phones over the last decade and a half. Reading these oral accounts against conceptual and empirical perspectives in the literature, I ask three questions. *First*, what has been the nature of engagement of Dalit castes with digital communication technologies such as mobile phones in peri-urban Bangalore? *Second*, have these technologies bypassed or been insufficiently harnessed by these historically disadvantaged social groups to overcome deprivation? And *third*, have they even assisted in the reinforcement of these exclusions for some of these groups?

To address these questions, we are called upon to familiarise ourselves with two broad sets of themes that support our enquiry. The first presents the meta-narratives that advocate the promise of information and communication technologies (ICTs) in developing societies, and especially for disadvantaged social groups; it also introduces the concept of the "information society" – a state of well-being that developing societies aspire to. This first set of themes is provided in this chapter. The second supportive theme assists in acquainting us with the economic condition of Dalits in contemporary urban economic settings; this will be treated in greater detail in Chapter 5. Besides the first set of defining themes, we also present in this chapter the methods employed by this study for empirical enquiry. In Chapter 5, we lay out the evidence from the empirical study, and elaborate on the underlying roots of the contemporary socio-technological outcome among subaltern communities in peri-urban south Bangalore, the location of our enquiry. Unpacking this convergence, this study builds a new perspective on the relationship between caste, ICTs, and development policy. I ultimately argue for a reexamination of the nature of digital divide, which is often encountered in development policy related to ICTs, and the social construction of the usage of technology that assists this divide. In the process, I also call for a deeper documentation and further research on the experience of the subaltern in the history of technological change in metropolitan India.[3] To put it in Anirudh Krishna's words, I wish to offer "a worm's eye view of digital life" (Krishna, 2017).

Information societies, pervading technologies, and mixed possibilities

Small devices, grand dreams

The setting of this study in peri-urban Bangalore is founded on the commonly held vision of the city being a central pivot in the information economy imagination, pictured by policy and popular perception as a fertile ground for spawning the archetypical information society with all the trappings of a "technological culture." Even before Independence in 1947, Bangalore had already made a mark on the science and technology (S&T) landscape in India, with both academic institutions and industries locating their establishments in Bangalore. Especially after the 1950s, and by the 1980s, Bangalore became one of the principal points along the locus of the S&T sector in India. Any long-time Bangalorean would affirm its already glowing reputation as a landmark city for excellence in electronics, instrumentation, space science, aviation, and so on, well before the information technology boom of the mid-1990s. Simultaneously, Bangalore from the 1990s demonstrated the features of a burgeoning metropolis that, on the one hand, held within it a gigantic informal industrial sector similar to that of South

Asia and Latin America but on the other hand, exerted to showcase itself as a city that shined with knowledge-based industries in the service sector (Heitzman, 2004), or presenting to the rest of India what is often referred to in the literature as a "post-industrial society," based on Daniel Bell's 1973 work. This was in seamless consonance with hopes around the information technology enabled services (ITES) sector as well, as seen in Chapter 1. In such a society – much like what Bangalore wished to project itself to be – the structure and operation of economy, society, and governance would be facilitated by service sectors empowered by digital high-speed communication technology. These dreams were not entirely baseless, and were drawn from the fact that ICTs have enjoyed, for a considerable period of time, a global reputation in their role in furthering positive development outcomes, particularly in advancing well-being among socio-historically and economically deprived groups. All over India, too, the language of development policy, economic transition, and social advancement was based on the ambition to upgrade from merely a "developing region" featuring in the lower extremes of United Nations Development Programme (UNDP) and World Bank data tables, towards a glittering "knowledge economy" as a beacon for other developing economies in South Asia and elsewhere. Solutions to address the challenges of socioeconomic and environmental well-being began gradually revolving around the deployment of ICTs and other new technologies.

Consequently, there were popular visions of an internet-based society where citizens would access information and public services through digital technologies. The emboldening of skill-learning among the literate and educated, the opening of access to new resources and networks among the traditionally marginalised, and so on, were expected to be automatic effects of ICTs, delivering discernible improvements in the lives and long-term livelihood prospects of individuals and households, such as for socio-historically deprived castes. The digital divide (which we will discuss soon) stared in the face of these national aspirations but was to be overcome successfully with "digital solutions" (Warschauer, 2003), rooted in both the general observation that ICTs play an overwhelmingly critical role in all aspects of contemporary economy and society, and the assumption that increased access to ICTs by deprived communities holds the promise to relocate them from historical marginalisation to seamless inclusion and participation in the contemporary era. The move away from traditional interpersonal face-to-face interactions for seeking socioeconomic resources, towards information access by clicking buttons or scrolling through screens would "flatten" society, and there would be free and fair access to public and private services regardless of the individual's caste, class, gender, or socio-political power. Information processing would be the core activity conditioning the productivity of production, distribution, consumption, and management of economic and political resources in such a setting (Kumar, 2006). In such an information society, digital technologies simply become an extension of one's bodily senses – of sight, sound, and speaking – and facilitate greater participation in economy and society to improve well-being. In other words, even one's selfhood and social relationships will be configured through and with digital technologies, resulting in greater atomisation of the individual (Lupton, 2015).[4] This flavour of a technological vision was actually taken quite seriously in the popular imagination, with digital technologies – particularly digital communication technologies, such as the mobile phone – percolating through every possible caste and class in India and imbuing every aspect of life, livelihood, and leisure.

The mobile phone dispersed through India – metropolitan, urban, and rural – like no other technological artefact did. Even at a global scale, it is claimed that nearly every human being is in principle within the range of a mobile phone signal (Lupton, 2015). From being an expensive communication instrument for the super-rich or as a device facilitating ostentation for the upwardly mobile in the late 1990s, the mobile phone emerged as *the* supreme personal communication device in India – which, Castells et al. (2007) declared, struck at the heart of human interpersonal activity, with immediate implications on information control and socio-political power, and the creation of a new form of public space contained within digital space. Mobile phones held the potential for broad information dissemination at low cost, across regions, and to people who were not reached out to by the conventional copper wire, orienting the economy on aggregate as increasingly dependent on information and democratic diffusion of knowledge as its central axis and lifeblood (World Bank, 1999). Job information, employment-related migration, social support, and so on, began to ride on the back of communication technologies (Qiu, 2009).

Jeffrey and Doron (2013) provide an extensive treatment of the entry and experience of mobile phone technologies in India, particularly among those populations that were at the middle and lower rungs of the socioeconomic ladder. They present the fact that for large sections of the Indian population, particularly in cities, the mobile phone (henceforth "the phone") had become the first major appliance bought by an individual or household, even before other appliances such as a television or refrigerator, or even a motorbike. The phone became a central characteristic and a ubiquitous artefact of the "new India," as opposed to the hitherto technologically excluded or isolated squalor of the poor and underclass. The phone allowed individuals, small businesses, informal workers, politicians, activists, friends, family, and practically *anyone* to transcend physical and social barriers in a country synonymous with hierarchies of every type imaginable. By 2010, the most basic versions of the device were within the financial reach of most of the population of India, as it cost much less than the fixed land phone. Everyone seemed to have use for it in their own way, and its fundamental operation was in easy grasp as it required only speaking and listening (unlike, say, the internet or personal computers, which required a skill and endowment set including command over the English language, familiarity with the interface, literacy, affordability of device, etc.). So many people who would have otherwise never had service-sector jobs were now customer-service executives in Nokia device support stores or Airtel mobile connection recharge centres. It aggravated Indian consumerism, expediting a sharp shift away from an otherwise very judicious and calculated paradigm of individual and family resource management. Mobile phones were envisioned to undermine old social structures of caste and gender, and were visualised to open information to even those who were supposed to be kept uninformed for the greater good of the traditional social structure. Lupton (2015) reiterates Manuel Castells, who had paved the way for acknowledging and giving due appreciation to the idea that these technologies were critical for contemporary social formation. Even within the home, structures and power relations were provoked by the presence of the phone, unlike in the case of so many other technologies thus far (Bell, 2006a, 2006b).

Doubts

However, whether at home, in local society, or in the larger economy, has the phone actually been successful in achieving all what has been alluded to it, or has it spawned new inequalities (Castells et al., 2007)? The first such new inequality that was born in our era of the phone is the *digital divide*, which is quite simply the division between those who have access to and regular engagement with the device (or for that matter, digital technologies in general), and those who do not. The digital divide is, first of all, about *access* (Kumar, 2006). Those on the unfortunate side of the divide are feared to have only remote access to the contemporary development experience, aggravating inequalities of a new variety and concentrating resources and skills to those better endowed (Warschauer, 2003; Saith and Vijayabaskar, 2008). In addition, agents of policy and development intervention, beginning with the state, have long held the belief that mere access to these devices can not only close digital divides but also address (if not solve) social problems in general (Warschauer, 2003).

An entire body of thought has revealed that bridging the access-based digital divide (i.e., the divide between simply owning and not owning a device), even completely, might not amount to anything significant enough to improving the conditions of society, as it may only aggravate social inequities further given its contradictory possibilities (Wajcman, 2004). For the underprivileged, simply holding and operating a phone may not facilitate a leap from the old world to a new world. This is simply because the digital and the social are not separate, but they are in fact mutually configured – a principal non-separation that the conventional understanding of the digital divide does not acknowledge (Halford and Savage, 2010). Extreme optimists of digital technology and mobile telephony have entirely missed the fact that digital divides mirror the wider patterns of social privilege and exclusion, and do not reverse entrenched inequities (Saith, 2008). Social inclusion in the contemporary economy is, even if strongly intended, not an automatic result of digital inclusion, as *digital* solutions may enable but will not mandatorily and inevitably undo and fix the *"analogue"* problems of society and economy (Warschauer, 2003; World Bank, 2016). Connectivity by itself is necessary but nowhere near sufficient; it certainly provides new opportunities and holds the potential to strengthen development trajectories, but this does not *ensure* that ICTs will be harnessed by those in need to move up their development trajectories, because the "analogue foundations" of society (referring to basic developmental concerns around education, health, gender, equity, etc.) will in all likelihood remain resilient and durable (World Bank, 2016).

Much of such naive thinking stems from a technocratic view of development and a notion of an information technology "revolution" as an edifice on which the "information society" is built (Webster, 2006). To reiterate from Chapter 1, technological determinism appears to be a paradigm of convenience, wherein social realities are mere nuisances to be either simply brushed under the carpet (Morozov, 2011), or imagined as self-disappearing as in the case of caste in urban India. In fact, Morozov (2011, 2013) provides strong and compelling arguments that technological determinism, assuming a "technology-fixes-all" line of thought and intervention, provides a sense of false hope and comfort. It may even aggravate a blindness to the previously mentioned technology–society nexus that is not simply a perspective but for the most part *fact*. Technological "solutionism," as critiqued by Morozov, ignores all the intricacies of deliberation and debate on how technologies – including digital communication technologies – and social structures and dynamics are not only related but co-constructed.

Uncertain possibilities

This is fundamentally because, as the literature has clearly established, the "stuff" of an innovation (be it a bicycle or a mobile phone) is not just science and engineering but also as much sociology and politics (Bijker et al., 1987; Bijker, 1997, 2010). There is sufficient evidence that identity and community are as relevant to virtual worlds as much as they are in the social world around us (Hammersley and Atkinson, 2007). The digital world is not flat (Stehr, 1994; Qiu, 2009), as ICTs such as mobile phones hold different *meanings* for different social groups who are unevenly empowered (Bijker et al., 1987; Bijker, 1997, 2010; Faulkner et al., 2010; Wajcman, 2004, 2010; Jeffrey and Doron, 2013; Ahmed et al., 2014; Lupton, 2015). That is, a technological artefact such as a mobile phone is interpreted by a social group through its own idiosyncratic perspective, to the extent that there may be heterogeneity in interpretation even within the same social group across geographical locales. Social groups may attribute subtle or overt meanings to the phone (or its usage) depending on their respective socio-hierarchical position and the perceived possibilities of livelihood improvement that it attaches to the device. Hence the access or usage of any technological artefact as a result becomes invariably embedded in highly context-specific social structures, social processes, power relations, interest groups, norms and values, and so on (Stehr, 1994; Warschauer, 2003; Bell, 2006a, 2006b; Guy and Karvonen, 2011). Sociopolitical variables significantly influence whether or not mobile phones will actually deliver their reputed positive outcomes in improving well-being.

Conventionally perceived a boon for the poor, digital communication technologies such as mobile phones also hold creative possibilities for *perpetuating* the exclusion of marginalised peoples from new livelihood-improving opportunities, improved social networks, and evolving social capital. This possibility is not remote or exceptional. Let us, for instance, consider social networks, which are the foremost source of livelihood information for the peri-urban poor (Schilderman, 2002). These networks can be either nourished or starved, which in turn strengthen or diminish the social capital of communities (as in the case of the Saliyars, discussed in Chapters 2 and 3). It is well known that even in the contemporary era, being a member of a dominant caste assures livelihood benefits that stem from the highly exclusionary nature of the internal social networks and the resultant strong social capital within these dominant castes.[5] The functioning of such dominant social groups (and their already rich social networks) get an added fillip through mobile phones, which in turn fuels new forms of exclusion against subaltern castes, as a result thwarting the attempts of the latter to improve their livelihoods. Besides in the formal economy in peri-urban Bangalore, in the informal economy in this region too, disadvantaged castes have been systematically excluded from valuable networks and opportunities. For instance, studies have found that most of the well-paid workforce in the ITES industry in Bangalore are those who come from well-off agricultural families or upper castes, and that there are strong spatial and social discontinuities in technological experiences within the economic region that we love to term as an "IT city" (Heitzman, 2004; Upadhya, 2007; Upadhya and Vasavi, 2006). Greater dependence on mobile phones might have actually exacerbated the economic exclusion of disadvantaged castes, especially in the informal sector that emerged as a spin-off from the ITES industry in this region – a reality remote from the information society imagination laid out at the beginning of this section.

Caste-based deprivation, which operated using a different and more brutish apparatus hitherto, may now be nurtured by employing mobile phones, which assist efficiently in pushing the already caste-based social capital into a very exclusive closed social network in digital space (see Prakash [2015a] for a detailed empirical treatment of this theme). And though the instant exchange of electronic messages using ICTs have, as studies have shown, created a "more intense sense of shared experience and of a shared social world" (Hammersley and Atkinson, 2007: 138), in reality they may or may not have really changed livelihood outcomes and well-being for the deprived in the long run. Revisiting Jeffrey and Doron (2013: 121), the mobile phone may have improved some conditions, but it did not reorder society. Hence, the durability of caste in peri-urban metropolitan India may have built an ugly nexus with the phone for furthering social exclusion, diametrically opposite to the well-intentioned vision of ICTs improving well-being and expanding livelihood opportunities by beneficially percolating through every aspect of life and livelihood.

Summing up

Very little of these complex dynamics are unappreciated or unaccounted for in the policy discourse in India. Popular and policy notions about the greater process of development often procreate from grandiose fantasies associated with advancing towards an "information society" or a "digitally empowered local economy," wherein old sociological variables such as caste are perceived to have aged and deteriorated, and are now too impotent to have any place or relevance in the crafting of socio-technical trajectories that lead us towards these ideals. The thriving forms of caste – both new and old – are given only passing mention (if mentioned at all) in the general policy discourse around ICTs in development in India. The sociological entity of caste has been popularly imagined to be "modernised" or "democratised," or even rendered "irrelevant" (Natrajan, 2012). This bent of understanding is reflected in the popular and policy discourse around the developmental role of ICTs in India, which have often been seriously inadequate in understanding these ground-level realities. Programmes such as Digital India or the National Digital Literacy Mission, which are indeed noble in their purpose, demonstrate a shallow understanding of the political nature of the relationship between technology and caste. Apart from an appreciable recent international policy recognition (very notably World Bank, 2016) that maximising digital dividends in a society requires working with the analogue foundations of that society, there has generally been in India a policy myopia of the deceptively complicated relationship between technology and society.

However, if phones may have become the newfound apparatus in creating oblique experiences for socio-historically deprived groups, have there been also responses from these groups to mitigate the previously named adverse effects? Do factors such as youth prod innovative usage of phones to renegotiate or rework power relations, and excise subordination from everyday life and work? Has the broad permeation of phones among the subaltern and the underclasses intensified the drive for social cohesion among them? Empirical realities that are presented in this study divulge that all these possibilities may not have materialised, and instead, the convergence between the sturdiness of caste, the usage of mobile phones, and credulous understanding of the digital divide has in most cases furthered the oblique experiences.

Site of empirical enquiry

Introducing the field site

The dualism that Upadhya and Vasavi (2006) portray in great detail is apparent even to an untrained eye who passes by on a fleeting drive through peri-urban south Bangalore and glances out at shining glass buildings, shopping arcades, and scintillating apartment complexes that exist alongside a complicated mix of peri-urban squalor, hasty agricultural transition, local dreary economies, and small tenements. This is a scenario that could easily fit into a classic textbook case of socioeconomic dualism. The area is called Choodasandra, enclosed by Sarjapur Road, Hosur Road, Hosa Road, and Central Jail Road, and it was inhabited for decades prior to the landing of the ITES Mayflower here in the late 1980s, gradually through the 1990s, and especially after 2000. With little regard for environmental sustainability and scant efforts to amalgamate with local economies and societies (which the older colonies of Bangalore-proper at least attempted), gigantic commercial and residential buildings and spinoff formal and informal economies emerged in the area at a breakneck pace over the last 15 years, delivering an experience of shock and awe to locals who had long inhabited these areas with their pastoral practices and inter-caste frictions. All along these roads, one easily notices stocks of three-storey or four-storey buildings with their characteristic bright painted (or unpainted) walls and crisscrossing metal staircases characterising the front view, inhabited by those who are evidently not part of the ITES sector but perpetually with mobile phone in hand.

According to the Karnataka District Census Handbook 2011 for Bangalore (from which all of the following data has been sourced), the area enclosed by Choodasandra, located within Anekal taluk, is around 160 hectares in expanse, with a population of under 3,000 (as of 2011 – this would have surely more than doubled with apartment complexes rising endlessly); there are two government schools in the region, and government health facilities are at least five kilometres away. There is no agricultural land here anymore, while there are a few garment and chemical factories around; nearly 80% of the population in this taluk engage in non-agricultural activities. Dalits constitute a quarter of the population in Choodasandra, slightly higher than the figure for Anekal taluk (around 21%), higher than the figure for Karnataka state (around 18%), and much higher than the proportion for Bangalore as a whole (around 12%). However, though the taluk as a whole boasts a literacy rate of around 80% (86% for men and 74% for women), which is equivalent to that of Bangalore as a whole, the literacy rate among the Scheduled Castes is around 12 percentage points lower.

I chose this area for empirical study given its very close proximity to the ITES belt, by virtue of which I assumed several new work opportunities (in the informal sector, of course) would have spawned. Its sizeable Dalit population is mostly concentrated in modestly designed apartment complexes earmarked to rehabilitate those dispossessed of their homes in slums in Bangalore city. Other castes in the region include the powerful Reddy and Lingayat communities, Gonigas, Bajantries, and Adi Karnatakas, residing in clearly segregated housing areas based on caste (Madsen, 2010). Caste conflicts around Choodasandra and its neighbouring area Junnasandra have been quite rampant even up until the 1980s, which Madsen (2010: 21) records:

> In 1984 the local big farmer had attacked the Dalit colony. The conflict that had built up over a long period, broke out because an agricultural labourer had refused to work due to a health problem. He was beaten up by the farmer's men. In a state of drunkenness some Dalit youth the following day shouted at a gang of Reddies from the nearby village Halanayakanahalli, who had to pass the centre of Junnasandra on their way to the bus. The following day, one of the Dalit boys was beaten up

by Reddy youths. Reacting on this infuriated Dalit boys retaliated by throwing stones after Reddies that had to pass the Dalit area on their way to Halanayakanahalli (see map 3). A few hours later, 300–400 men in 4 tractors arrived in the village and beat up all the Dalits, who had not gone to work in the morning. They cut the arms, legs and faces of the Dalits with coconut and harvesting knives. Those who could walk fled the village screaming and after some hours gathered in front of the parliament building in Bangalore, demanding help to the wounded and police protection of their village.

Three cohorts

In order to address the objectives of this study, I identified three different cohorts for empirical study within Dalits in Choodasandra. Cohort A, households, is located in a traditionally settled region of historically deprived Dalit castes, which immediately display deprivation (in variables such as quality of housing, access to water, access to quality education and health services, access to upward economic mobility, and so on). Cohort B, at individual and group level, comprises those individuals from within these Dalit castes who have come from a peri-urban and deprived background similar to Cohort A but have experienced dramatic upward mobility over the last decade to the extent that they manage or even own businesses of various sizes in Bangalore city. Cohort C includes about a dozen individuals and organisations among the vast network of civil society workers and Dalit activists in Bangalore, most of whom have also originated from the setting of Cohort A but who operate in the peri-urban and metropolitan setting at the core of socio-political issues that characterise Cohort A.

Let us look at each setting in some detail. The selected area of study for Cohort A – two housing colonies at Choodasandra – has been picked by virtue of the context it provides to study the engagement of digital communication technologies and historically deprived castes in terms of livelihood improvement. The area is located in proximity to Electronics City, one of Bangalore's major ITES enclaves. The entire region is substantially populated with demarcated dwelling areas for Dalit castes, which have historically suffered caste-based atrocities in the hands of the upper castes as well as wealthy and prosperous land-owning castes that dominate this region. Offices of several organisations (such as the Dalita Sangharshana Sene, Dalita Sangharsha Samiti (DSS), and Dalit Panthers of India (DPI), among many others) catering to the welfare of Dalit communities can be spotted in the area. I focused on two housing clusters with 96 families in total and interviewed members of 71 households (of a total of 214 individuals, each family comprising of two to four individuals). The other 25 households were not present across all three months of the fieldwork, and I assumed that household members had migrated, most likely for employment. Two housing colonies with nearly 100 families in total (each family comprising of two to four individuals) were identified, of which nearly 70 households were met with personally, and interviewed.[6] One colony was clearly older than the other, but the homes were generally owned by families of these Dalit castes, and often rented out to other families (which were not necessary of Dalit castes but economically highly deprived). The colony was characterised by concrete houses, access to regular water supply and sanitation facilities within the house but other issues such as visible malnutrition in children, unemployment among the young, illiteracy or deficient school education, and a general sense of precariousness in livelihood, was all too apparent. These were some characteristics that were deliberately sought, to avoid those households and settlements that are abysmally poor. This study wishes to understand those households and families who may benefit from living conditions that are generally above the average for a similar caste status elsewhere in India but for whom intense engagement with mobile phones may not have translated into positive outcomes by virtue of their caste networks or interpretation of the technological artefact.

Cohort B, on the other hand, was located not at one single site but situated within Bangalore city, consisting of individuals who originated from Dalit castes especially from peri-urban Bangalore. These individuals, who have vivid memories of living in conditions equal or worse than those of Cohort A, now proudly state that they have risen high, to the extent that allegedly caste "didn't matter" to them anymore (as one respondent alleged). These were individuals, all male, who own businesses large and small ranging from liquefied petroleum gas (LPG) distribution outlets, to electrical shops, to small firms, right up to one well-recognised and felicitated individual who has featured as a "Dalit millionaire" in various popular media pieces (as well as in Kapur et al., 2014). All these individuals now reside and operate within Bangalore city, and have built a strong socioeconomic network among themselves.

Cohort C, again not in one locale, is a sample of civil society workers and activists who by nature of their operations brings them to the core of locations and experiences where the durability of caste displays its most open and brutish form. These are individuals from a spectrum of organisations that include the Dalita Sangharshana Sene, Dalita Sangharsha Samiti, Dalit Panthers of India, Dalita Samara Sene, DSS Ambedkar Vedi, and Karnataka Mahila Dalit Vedike, among others. Most are located in peri-urban Bangalore around Electronics City, while some individuals and organisations operate from within the city.

To sum up, this study spans its empirical enquiry across three cohorts: (A) households of Dalit families in a socioeconomically deprived neighbourhood in peri-urban south Bangalore; (B) individuals engaged in commercial or entrepreneurial ventures, who belong to Dalit castes and who have originated from households as in Cohort A but have experienced tremendous upward economic mobility over the last decade; and (C) civil society activists who often also hail from these Dalit castes and settings as in Cohort A but who have unique experiences with ICTs, compared to Cohorts A and B and who are at the forefront of challenging historical deprivation. The nature of each cohort ought to immediately reveal its idiosyncratic interpretation and usage of the mobile phone. Methods of enquiry on these idiosyncrasies also differed with the highly divergent nature of these cohorts.

Methods

The entire data collection process aimed at eliciting primary oral information on what role the mobile phone has played in livelihood improvement over the last two decades, for access to valuable networks and other livelihood generation opportunities. Documenting *micro life-histories* of the cultural, sociological, and political engagements with the mobile phone has been the central method of this study, for each cohort.

Oral life-histories are a type of in-depth interview at an individual level, with focus on personal experiences across a lifetime, and with mostly qualitative information elicited in an interactive manner by the researcher, by which the participants' subjectivities are located in a broad social and cultural context (Porta, 2014). This method is valuable for explanatory accounts from not hundreds but at maximum a few dozen individuals (Porta, 2014). Memories and personal commentaries form the information of life-histories, with information saturation being the point of closure of the interview (Bertaux, 1981; Ritchie, 2003). As a method, it brings out subjectivity, variation, and degree of disadvantage among especially subaltern groups, this heterogeneity being very valuable to the larger story. In fact, the literature on oral life-histories (for instance Prins, 2001; Ritchie, 2003; Janesick, 2014; Bosi and Reiter, 2014) forcefully argues that this methodology provides a historical presence to those whose views and values have long been disenfranchised by "history from above." This could well include the history of technological change in metropolitan India. This is the epistemological basis of the principal methodology chosen for our study, though in this study it may not seek highly detailed and lengthy oral life-histories common to ethnographic studies of a larger scale. Instead, we elicited what I term as micro life-histories, which shares the objective of the life-histories method in general but is a much-shortened version spanning a brief period in the respondent's life, and focused around a particular context – the experience with the mobile phone. The entire data collection of recording oral information was conducted in Kannada, which was translated and transcribed by an assistant.[7]

For Cohorts B and C – the businessmen and activists – lengthy, open-ended interviews aiming at constructing micro life-histories were employed to find out what their experience was, over the last 15 years, with the mobile phone. Eleven individuals were interviewed in Cohort B, and in addition a focus group discussion was conducted with all 11 together, under the auspices of the Bangalore chapter of the Dalit Indian Chambers of Commerce and Industry (DICCI). In fact, the DICCI was instrumental in gaining access to the entire cohort. We elicited information on their backgrounds, their narratives on the economic situation within and around Dalit castes in especially the peri-urban regions of Bangalore, how they individually moved out of these hard circumstances, and how

91

the mobile phone entered their lives and influenced their upward trajectories over the last 15 years.

For Cohort C, 13 individuals from different activist organisations (selected by snowball sampling beginning from the first office in the neighbourhood of Cohort A) were interviewed on their experiences with the device. Given the intense engagement that these individuals have with the phone, it was interesting to collect information on how they perceived the device not only in the obvious sense of faster and more instantaneous communication for collectivisation and social mobilisation but as to whether the mobile phone has brought new challenges to activism.

For Cohort A, therefore, the objective was to elicit micro life-histories of their cultural, psychological, sociological and, most importantly, political engagement with the mobile phone. The whole of Cohort A was interviewed by going door to door across both colonies, and building rapport with the family before interviewing them. We built rapport by asking them whether they knew programmes such as Digital India, or briefly introducing it to them if they were unfamiliar with it (which was the case among almost all households). The short oral histories of engagement with the mobile phone were elicited by means of a questionnaire with five categories of questions.

1   *On Encountering the Phone*: what kind of a mobile phone do you have? For how long have you had this? If this isn't your first phone, what kind was that, and why did you change it? Do you wish to upgrade your phone now, and why?

2   *On one's Personal Relationship with the Phone*: how much time do you spend on this? Do you have internet access on it? Does it help your job? Does it help maintain better contact with family and friends? Who else in your household owns a phone, and what do they mostly use it for?

3   *On the Role of the Phone in the Job*: how does the phone help you in your everyday work? Do you see how phones are helping so many people better their lives by gain new professional contacts and improving their jobs, or getting new jobs? Do you think you're also going to benefit this way sometime?

4   *On Caste Solidarity and the Phone*: do you maintain contact with the local DSS/DPI (Dalita Sangharsha Samiti/Dalit Panthers of India) leaders or any police officer or MLA (member of legislative assembly)? If you're using platforms such as WhatsApp or Facebook, are you part of caste-solidarity groups? Do you watch news and updates about events and programmes (religious or civic) through these platforms? Do you share caste-related issues or problems, or civic issues with your contacts through your phone?

5   Does the phone make you feel more "liberated" compared to your parents or grandparents, and how? Or does it open up negative experiences too?

In Chapter 5, we proceed to the empirical findings of these micro life-histories and interviews, and a detailed discussion of their interpretation.

Notes

1 An earlier and much shorter version of this chapter has appeared as A. Kamath (2018), "'Untouchable' Cellphones? Old Caste Exclusions and New Digital Divides in Peri-Urban Bangalore," *Critical Asian Studies*, 50(3): 375–394. Used with permission of Taylor & Francis Ltd, www.tandfonline.com, DOI: 10.1080/14672715.2018.1479192.
2 Here, Dalit or formerly untouchable castes, who have traditionally been placed at the very bottom of the Indian caste system and for centuries have been subject to severe deprivation and often even intimidation and violence. The term Dalit, according to Prakash (2015a, 2015b) has come to signify the political identity of people who have been called scheduled castes in the official vocabulary of the colonial and post-colonial state.
3 One important work in this regard in the context of the United States is Sinclair (2004).
4 Webster (2006: 7) points out a separation between those who endorse the idea of an information society and those who regard informatisation as the continuation of pre-established relations. There are on the one hand those who proclaim a new society altogether, who may be classified as theorists of post-industrialism (Daniel Bell), postmodernism (Jean Baudrillard, Mark Poster, Paul Virilio), flexible specialisation (Michael Piore and Charles Sabel, Larry Hirschhorn), and the informational mode of development (Manuel Castells). On the other hand, there are those who do not support the idea of an overhaul or evolution, but emphasise continuities with long-established principles of political economy, such as theorists of neo-Marxism (Herbert Schiller), regulation theory (Michel Aglietta, Alain Lipietz), flexible accumulation (David Harvey), reflexive modernisation (Anthony Giddens), and the public sphere (Jurgen Habermas, Nicholas Garnham).
5 The contemporary evidence on this includes Thorat and Newman (2010), Thorat et al. (2010), Das (2010), Thorat and Attewell (2007), Harriss-White (2004), Harriss-White et al. (2014), Kapur et al. (2014), Thorat and Sabharwal (2015), Verma (2015), and Prakash (2015a, 2015b). The durability of caste in metropolitan India will be treated with greater elaboration in Chapter 5.
6 Vinay Kumar's assistance during fieldwork was invaluable.
7 Again, the assistance of Vinay Kumar must be acknowledged here.

References

Ahmed, H., Qureshi, O.M., and Khan, A.A. (2014) 'Reviving a Ghost in the History of Technology: The Social Construction of the Recumbent Bicycle', *Social Studies of Science*, 45(1): 130–136.
Bell, G. (2006a) 'Satu Kelugara, Satu Komputer (One Home, One Computer): Cultural Accounts of ICTs in South and South East Asia', *Design Issues*, 22(2): 35–55.
Bell, G. (2006b) 'The Age of the Thumb: A Cultural Reading of Mobile Technologies from Asia', *Knowledge, Technology, and Policy*, 19(2): 41–57.
Bertaux, D. (1981) 'From the Life-History Approach to the Transformation of Sociological Practice', in Bertaux (ed.) *Biography and Society: The Life History Approach in the Social Sciences*, Sage, London.
Bijker, W.E. (1997) *Of Bicycles, Bakelites, and Bulbs*, The MIT Press, Cambridge, MA.
Bijker, W.E. (2010) 'How Is Technology Made? That Is the Question!', *Cambridge Journal of Economics*, 34(1): 63–76.
Bijker, W.E., Hughes, T.P., and Pinch, T. (1987) *The Social Construction of Technological Systems*, The MIT Press, Cambridge, MA.

Bosi, L., and Reiter, H. (2014) 'Historical Methodologies: Archival Research and Oral History Social Movement Research', in Porta (ed.) *Methodological Practices in Social Movement Research*, Oxford University Press, Oxford, UK.

Castells, M., Fernandez-Ardevol, M., Qiu, J.L., and Sey, A. (2007) *Mobile Communication and Society: A Global Perspective*, The MIT Press, Cambridge, MA and London.

Das, M.B. (2010) 'Minority Status and Labour Market Outcomes: Does India Have Minority Enclaves?', in Thorat and Newman (eds.) *Blocked by Caste: Economic Discrimination in Modern India*, Oxford University Press, New Delhi.

Faulkner, P., Lawson, C., and Runde, J. (2010) 'Theorising Technology', *Cambridge Journal of Economics*, 34(1): 1–16.

Guy, S., and Karvonen, A. (2011) 'Using Sociotechnical Methods: Researching Human-Technological Dynamics in the City', in Mason and Dale (eds.) *Understanding Social Research: Thinking Creatively about Method*, Sage, London.

Halford, S., and Savage, M. (2010) 'Reconceptualising Digital Social Inequality', *Information, Communication, and Society*, 13(7): 937–955.

Hammersley, M., and Atkinson, P. (2007) *Ethnography*, Third Edition, Routledge, London and New York.

Harriss-White, B. (2004) *India Working: Essays on Society and Economy*, Contemporary South Asia, Foundation Books, New Delhi, and Cambridge University Press, UK.

Harriss-White, B., Basile, E., Dixit, A., Joddar, P., Prakash, A., and Vidyarthee, K. (2014) *Dalits and Adivasis in India's Business Economy: Three Essays and an Atlas*, Three Essays Collective, India.

Heitzman, J. (2004) *Network City: Planning the Information Society in Bangalore*, Oxford University Press, New Delhi.

Janesick, V.J. (2014) 'Oral History Interviewing: Issues and Possibilities', in Leavy (ed.) *The Oxford Handbook of Qualitative Research*, Oxford University Press, Oxford, UK.

Jeffrey, R., and Doron, A. (2013) *Cell Phone Nation*, Hachette India, New Delhi.

Kapur, D., Babu, D.S., and Prasad, C.B. (2014) *Defying the Odds: The Rise of Dalit Entrepreneurs*, Penguin Random House India.

Krishna, A. (2017) *The Broken Ladder: The Paradox and the Potential of India's One Billion*, Penguin Random House India.

Kumar, D. (2006) *Information Technology and Social Change*, Rawat Publications, Jaipur.

Lupton, D. (2015) *Digital Sociology*, Routledge, Oxon and New York.

Madsen, A.M. (2010) *Dalits in South India: Stuck at the Bottom, or Moving Upward?*, Social Skriftserie No. 10, Department of Social Work, University of Aarhus, Denmark.

Morozov, E. (2011) *The Net Delusion: How Not to Liberate the World*, Penguin, London.

Morozov, E. (2013) *To Save Everything, Click Here: Technology, Solutionism, and the Urge to Fix Problems That Don't Exist*, Penguin, London.

Natrajan, B. (2012) *The Culturization of Caste in India: Identity and Inequality in a Multicultural Age*, Routledge, Oxon and New York.

Porta, D.D. (2014) 'Life Histories', in Porta (ed.) *Methodological Practices in Social Movement Research*, Oxford University Press, Oxford, UK.

Prakash, A. (2015a) *Dalit Capital: State, Markets and Civil Society in Urban India*, Routledge, New Delhi.

Prakash, A. (2015b) 'Dalit Capital: Discrimination, Unfavourable Inclusion, and Intersectionality', in Verma (ed.) *Unequal Worlds: Discrimination and Social Inequality in Modern India*, Oxford University Press, New Delhi.

Prins, G. (2001) 'Oral History', in Burke (ed.) *New Perspectives on Historical Writing*, The Pennsylvania State University Press.

Qiu, J.L. (2009) *Working-Class Network Society: Communication Technology and the Information Have-Less in Urban China*, The MIT Press, Cambridge, MA and London.

Ritchie, D.A. (2003) *Doing Oral History: A Practical Guide*, Second Edition, Oxford University Press, New York.

Saith, A. (2008) 'ICTs and Poverty Alleviation: Hope or Hype?', in Saith, Vijayabaskar, and Gayathri (eds.) *ICTs and Indian Social Change: Diffusion, Poverty, and Governance*, Sage, New Delhi.

Saith, A., and Vijayabaskar, M. (2008) 'Introduction', in Saith, Vijayabaskar, and Gayathri (eds.) *ICTs and Indian Social Change: Diffusion, Poverty, and Governance*, Sage, New Delhi.

Schilderman, T. (2002) 'Strengthening the Knowledge and Information Systems of the Urban Poor', ITDG, DFID, UK.

Sinclair, B. (2004) *Technology and the African-American Experience: Needs and Opportunities for Study*, The MIT Press, Cambridge, MA and London, UK.

Stehr, N. (1994) *Knowledge Societies*, Sage, Thousand Oaks, New Delhi, and London.

Thorat, S., and Attewell, P. (2007) 'The Legacy of Social Exclusion: A Correspondence Study of Job Discrimination in India's Urban Private Sector', *Economic and Political Weekly*, 42(41): 4141–4145.

Thorat, S., Kundu, D., and Sadana, N. (2010) 'Caste and Ownership of Private Enterprises: Consequences of Denial of Property Rights', in Thorat and Newman (eds.) *Blocked by Caste: Economic Discrimination in Modern India*, Oxford University Press, New Delhi.

Thorat, S., and Newman, K.S. (2010) 'Introduction: Economic Discrimination, Concepts, Consequences, and Remedies', in Thorat and Newman (eds.) *Blocked by Caste: Economic Discrimination in Modern India*, Oxford University Press, New Delhi.

Thorat, S., and Sabharwal, N.S. (2010) 'Social Exclusion and Poverty: Linkages, Consequences, and Policies', in Verma (ed.) *Unequal Worlds: Discrimination and Social Inequality in Modern India*, Oxford University Press, New Delhi.

Upadhya, C. (2007) 'Employment, Exclusion, and Merit in the Indian IT Industry', *Economic and Political Weekly*, 42(20): 1863–1868.

Upadhya, C., and Vasavi, A.R. (2006) 'Work, Culture, and Sociality in the Indian IT Industry: A Sociological Study', Report Submitted to the Indo-Dutch Programme for Alternatives in Development, National Institute of Advanced Study (NIAS), Bangalore.

Verma, V. (2015) *Unequal Worlds: Discrimination and Social Inequality in Modern India*, Oxford University Press, New Delhi.

Wajcman, J. (2004) *Techno Feminism*, Polity Press, Cambridge, UK.

Wajcman, J. (2010) 'Feminist Theories of Technology', *Cambridge Journal of Economics*, 34(1): 143–152.

Warschauer, M. (2003) *Technology and Social Inclusion: Rethinking the Digital Divide*, The MIT Press, Cambridge, MA and London, UK.

Webster, F. (2006) *Theories of the Information Society*, Third Edition, Routledge, London and New York.

World Bank (1999) *World Development Report 1998/99: Knowledge for Development*, The World Bank and Oxford University Press, Washington, DC.

World Bank (2016) *World Development Report 2016: Digital Dividends*, The World Bank, Washington, DC.

5

UNPACKING A CONVERGENCE AND EXPLORING NEW DIGITAL DIVIDES[1]

In this chapter, I present the findings from the empirical study across the three cohorts, and bring together evidence from the literature and theoretical perspectives to expose how the contemporary socio-technological outcome is a result of a convergence of three major actualities.

Findings

Cohort A

Every single household we surveyed in Cohort A owned a mobile phone. Most individuals in these households had owned a device for many years, which they had upgraded over the years. For many, it was the first electronic device they had ever bought. This is reminiscent to Jeffrey and Doron (2013), who explained how mobile phones are, in the contemporary Indian economy, the first devices bought by individuals in deprived social groups, acquired by households sometimes even before a vehicle such as a bicycle. However, it was doubtless that all individuals within a household did not own a similar device; often not everyone in the household individually owned one, as is the case with households of even a modest economic strata in India. Ownership of phones (and sophistication of the device) within households in Cohort A was generally unequal. Nearly all male individuals owned and operated mobile phones. This was not the case with all women in this cohort, who either owned a more rudimentary model of a phone (almost never smartphones) as compared to the male or younger members of their family, or would state that they did not *need* a phone at all, which was interesting to note given the ubiquity of the mobile phone in India regardless of a person's "need" for it. Young men and women (under the age of 25), however, displayed equal ownership (of 100 percent for each), and also possessed smartphones as compared to older individuals. Among the elderly (above 55 years of age), this gender gap was far starker, with elderly women for the most part not owning a device at all, and elderly men owning only a traditional mobile phone, not a touchscreen smartphone. Characteristics and information on ownership of the phone are presented in Table 5.1.

Table 5.1 Characteristics of Cohort A

Age group	Men	Women	Total
Aged under 25	51 (0)	13 (0)	64
Aged 25 to 55	78 (1)	30 (1)	108
Aged over 55	31 (4)	11 (9)	42
Total	**160**	**54**	**214**

Note: Figures in parentheses indicate the number of individuals who did not own a phone.

Source: Author's fieldwork

It was observed that almost everyone in this cohort used mobile phones primarily as communication devices, such as for keeping in contact with their employers or superiors (for the intermittent periods during which they actually were employed, nearly always in an informal job). Individuals in this cohort generally perceived a phone as anything beyond a device that makes communication convenient. There was nothing more to it, just an artefact to speak to people more conveniently, and while on the move. In several households that we visited during the empirical work, mobile phones were heard blaring out music for the benefit of the entire household. A series of mobile phone recharge plugs and their tangled wires dangling across the wall were a standard sight in every household. Often, we noticed families peering into these devices, enjoying a film song or a soap opera. There was nothing heard from our respondents in terms of how mobile phones had "alleviated" them from historical exclusion, or how it was now the most streamlined, unobstructable medium of collective association or mobilisation.

The younger men spent a far greater length of time on their devices, during most of the daytime and well into the night. Their phones played a central role in entertainment, social media communication, and for informal conversation among themselves. Sharing content sourced and circulated through mobile phone internet was the major activity during the day. While this was a serious issue of disappointment and irritation between parents and elderly on the one side and younger male family members on the other, most respondents on either side of this argument were convinced nonetheless that this form of mobile phone usage was the "reality" of these devices, which had to be accepted objectively. An elderly respondent lamented:

> we were never like this as youngsters. Our parents wouldn't allow us to even look at a girl! Now these useless boys even taunt young girls with bad pictures and messages. They keep giggling showing each other things, spending the whole day under that bus stop or near the auto-rickshaw stand. But we have to accept the modern day reality.

The young men referred to here, who in fact characterised the young men of our cohort in general, were rarely employed and generally had little education. They

97

spent much of their time in small groups with similar young men, sharing among themselves amusing or vulgar content on their devices. Exceptions to this were very few: in fact, only three individuals among young men were found to be regularly at work. Many elderly interviewees lamented that if they had had access to these devices during their youth, "things would have been much better" for them, assuming that their long-term livelihoods might have improved. Most people in this cohort rarely kept in touch with local civil society or political groups, and very few maintained phone numbers of local police or other local political actors who might be helpful during emergencies. They shrugged or complained:

> what's the use? These politicians and activists are around only for their own individual and political organisation interests. I have no idea who the influential people around here are, and it doesn't matter anyway.
>
> if I need something, I can help myself without approaching those self-ish people.
>
> I don't really recall any such messages on WhatsApp or Facebook, collecting everyone together for election cards or Dalit political rallies.

For many, the initial fascination of owning a digital device waned after a few weeks of purchase, after which it only became a convenience, never once imagined as a tool that would permit them access or inclusion to economic networks of the upper or dominant castes of the region. Upon enquiring about this possibility, many responded with blank stares, often accompanied by the simple yet powerful statement: "all this [upward mobility] is not for us."

A minuscule number (only two individuals out of all the Dalit households surveyed, and both under the age of 40) have some functional grasp on the English language. This was a significant enquiry in our empirical study, as it was soon realised that only these individuals were able to access and succeed in operating simple employment-related apps or a rudimentary job search online. The act of searching online for job openings via portals or apps does not require high technical qualifications, and is hence quite popular among aspiring lower-middle-class and working-class individuals in peri-urban Bangalore. Several respondents commented that members of the dominant and economically well-to-do castes in the region (such as the Reddys, Gowdas, Lingayats, and other Kshatriya and Brahmin castes) had their monetary and political fortunes skyrocket after the entry of the mobile phone era, not only because of their immense social and financial capital – both of which were historically established for these communities in this locale – but also out of a moderately good familiarity with the English language. Unlike for most of Cohort A, the younger individuals of these well-to-do castes, and now their children too, are sent to the nearest school presenting English as the medium of instruction (regardless of the quality of English spoken at these schools, or for that matter the quality of teaching in general), which naturally costs much more than public government-run schools. Public schooling in the region, not very different in quality from most of the state, is abysmal in terms of infrastructure and teacher quality, and English is taught as a peripheral language. The younger individuals in Cohort A, who have just entered their 20s, were beneficiaries of public schooling and not English-language education, which they complain was their second primordial deprivation, after their caste status.

The worst affected in Cohort A were actually found to be the middle-aged – for whom opportunities remain inaccessible, even after owning and operating mobile phones for over a decade. Their own social contacts are often battling financial and economic deprivation, and access to social networks of more well-to-do groups is beyond their horizon because of caste prejudices in informal interpersonal or economic networks. Personal interaction with members of dominant castes in the region was formerly severely restricted, or permitted only in a subordinate capacity. Respondents reported how they were often met with a cold shoulder, subtle insults, or occasionally even casteist slurs when they sought opportunities to expand their livelihoods in the region. This quiet but pervasive discrimination is said to continue today even in digital space, as useful and resourceful contacts through phones are exclusively shared only within these dominant groups. Some respondents commented:

> if we want to start a paint and hardware shop, who's going to give us the money? The only people who can give us credit for the very little collateral we have are moneylenders who will surely cheat us. And besides, it is impossible to get those contacts who can give you better jobs or better suppliers to begin a small business. These contacts are only with the well-to-do landowners and politically powerful here. We are met with indifference, or instead they tell us to become painters rather than opening paint shops. This clearly shows how they just don't want us to expand and grow.

Caste discrimination in economic spaces, and now even in digital spaces, are a lived reality of members of this cohort. Most individuals spend most years of their lives shuttling between a variety of informal and petty jobs that may include working as security guards, construction labour, domestic labour in the nearby apartments, small-time plumbers and cleaners, and at very maximum, auto-rickshaw drivers. Only one individual worked as a driver associated with Uber, while another was a school bus driver. To move to anything beyond these jobs, or to climb up a ladder slowly and steadily, was reportedly out of bounds for nearly all individuals here. The mobile phone appeared to be a passive companion in their lives, allowing them nothing more than simple interpersonal communication or entertainment.

Cohort B

In a completely different experience, all 11 individuals in Cohort B not only owned mobile phones but also stated that their phones were not a passive companion – rather, they were one of the essential keystones to their financial and social upward mobility over the last decade or more. What was especially different from Cohort A was that individuals in Cohort B believed their phone was essential not only for communication in their existing (formal or informal) job but also to facilitate the expansion of their existing business and livelihood opportunities – a virtue of this technology often highlighted in the literature and policy discourse around ICTs. Cohort B members believed their device did exactly this, i.e., to expand their interpersonal networks and to allow them to broaden their capabilities and fuel their upward mobility. A mobile phone, they said, assisted in their migration to the interior of the city from the peri-urban and rural locations they originally hailed from, and also allowed them to get in touch with sources of capital and crucial business information within the city. Though there were small informal businesses among Dalits in metropolitan and peri-urban Bangalore well before mobile phones dominated the economic scene, this tech-nology expedited the enhancement of entrepreneurship possibilities, even if at a modest level, as well as enabled the diversification of business networks. However, the most interesting issue that was brought up during the interviews and focus group discussions with members of this cohort was that mobile phones allowed for a certain extent of caste-anonymity, or even "liberation" from caste identity. One respondent from this cohort stated:

> we can easily gain new contacts of resourceful people in the city, even from higher castes, through this phone . . . after all [laughing], they can't hear my caste when I speak to them. In fact, not only across Bangalore . . . I can source new vendors even in towns just an hour or two away with the phone. . . . I don't know their caste, as much as they don't realise mine . . . who cares, all we need mutually is a good price and quality service, not caste.

Hence, respondents explained how they could gain new contacts within the city across castes through their phones, as caste "could not be heard through the device." They claimed that the only things that mattered across phone conver-sations were the quality of the service or product, a competitive price, speedy delivery of the product or service and, most importantly, trust. Though the last element is often tied to caste-based social capital (as illustrated extensively in the earlier chapters on the Saliyars), assurance in the other elements ensured the building of trust. But the Cohort B respondents also characterised members of Cohort A in highly derogatory ways, for example:

> they're just lazy. What does it cost to leave that place and come to city? What do they lose? After all, the only thing they leave behind are some freebies that some politicians gave them or their ancestors to get some votes for momentary political gain. No use staying there and constantly

crying, or constantly demanding things from the government. Just come here [to the city].

So according to this respondent, all that the young and able in Cohort A had to do, "obviously," was to migrate into the city and use the immense power of this technology to expand their networks, and eventually their possibilities of upward socioeconomic mobility, by trying out new jobs:

> look at that fellow from North India selling street food. Does anyone ask him his caste here in city? Does anyone care for anything apart from the taste of his cooking? See? There's no point depending on government reservations in jobs or some political activity. Especially political activity is a waste of time for the youth and for the enterprising. The phone helps you get many things here, without bothering about caste.

Cohort C

Members of Cohort C revealed that it might be far-fetched to claim that an individual, particularly a Dalit, can entirely conceal one's caste identity with a phone (or that it was irrelevant altogether, again especially for Dalits). There is little doubt, they argued, that caste is still highly durable and functioning in a significant capacity in the city, especially for economic networks and for progress. Many small businesses still prefer "their own people" to work with or employ, and though price and quality of service were central, most products and services of a higher standard were off-limits for these Dalit businessmen. Many respondents in Cohort C downplayed claims by Cohort B, arguing further that educational qualification and command over English to deal with more well-to-do customers, as well as a bit of local political clout, still mattered. However, simultaneously, many respondents in Cohort C revealed that mobile phones have indeed opened a virtual space for Dalit activists to operate within, where the critical importance of caste is marginally eroded as well as explicitly presented at the same time, depending on the need at hand. While Cohort B respondents might have benefited from the marginal erosion of caste identity in some quarters of their activity, individuals and organisations in Cohort C explained how caste was the very axis around which they developed intervention strategies and exercised agency, using phones as communication and mobilisation devices. Some activists in Cohort C boasted about how via social media platforms they were able to collect people and families together around the purpose of caste solidarity and increased political participation. One respondent emphatically stated:

> just a few months ago, everyone was brought together for camps that I organised to apply for voter identification cards and ration cards [to access public distribution systems for food and fuel]. There are so many

government subsidies for our community! And every other month, there are religious festival related celebrations for which we can all get together. All possible with WhatsApp.

However, others within Cohort C stated that there was inertia among individuals and families in Cohort A to utilise their phones for building cohesion among themselves and with local activists. This must be read against the common finding in Cohort A, about how contact with local activists was actually nonexistent, and that this sort of claim, quoted previously, was just a "lie." In fact, during fieldwork, we were provided this statement quite earnestly by a prominent ground-level Dalit activist in Choodasandra, before we began fieldwork with Cohort A. This statement was repeated often to respondents in Cohort A to elicit their response to this claim, hoping to hear some positive utilisation of the phone, at least for activism and collective association with local leaders. However, for the most part we received only negative responses, that this was just fiction, and that most people in Cohort A didn't recall any such messages on WhatsApp about festivals or political mobilisation. As quoted earlier, the response to this was even to the extent of calling some named members of Cohort C either as "unheard of" or even "selfish"!

Let us move to activists in Cohort C outside of Choodasandra. Most respondents of Cohort C who operated in Bangalore city proper stated that their social movements were alive and functioning even before the advent of widespread cellphone usage, and that the technology provided them an extra boost for collective mobilisation and quick communication across distant geographical locales. Several activists who are prolific writers on the cause of Dalit emancipation in both English and Kannada languages have found phones and ICTs in general a greatly liberating venue to escape not from caste as such (as claimed by some in Cohort B) but to find a digital space that is an alternative from regular press and media coverage of caste questions and issues. One prominent activist from this cohort stated:

> as a development journalist, things are so much easier for me now. Back in the day, I was restricted by newspaper editors and vernacular magazines from publishing my work. I don't need to beg them anymore. I have my regular blog, followed by thousands who from their part also didn't have access to reading this sort of thought. My fellow comrades and I write for so many online venues with no restrictions. No more gatekeepers!

There is today in Bangalore, as elsewhere, a plethora of online forums, blogs, pages within Facebook, small newspapers, and other venues for Dalit writing, spawned mostly after the entry of the mobile phone (and other ICTs). Interviews with this cohort revealed that most openings for Dalit activism were not through traditional pamphlet distribution or independent newspapers – not even through street theatre or marches – but through online avenues and rapid circulation (or as colloquially expressed, "viral") of short messages, pictures, posters, and symbols of Dalit solidarity through platforms such as WhatsApp or Facebook.[2]

However, the converse experience has been that those who perpetuate caste-based subordination ideologically, spread false information targeting Dalits, or simply spew crude hatred, have also used this new digital space to operate and expand their intent:

> the saffron brigade [the Hindu right-wing] now circulate all sorts of stuff insulting our food habits, such as eating meat or even beef, which is an insult to much more than just food preferences. At times, they even openly encourage violence against those who go against the traditional caste order, such as by intermarrying or indulging in political activity. Such individuals and groups now have access to these phones and the power that comes along with them.

An instance was cited by two respondents (interviewed together) in Cohort C of how, around a lake in peri-urban Bangalore, a small group of individuals noticed a few Dalit young men fishing in the lake. Infuriated, they sent messages to attend to this pressing issue of Dalits fishing in a lake from which they were traditionally barred, and in no time a gathering of armed men arrived on the scene to fiercely intimidate and warn these men of dire consequences if they repeated this abominable act. The Dalit boys, apparently, were unable to message for help because the rival group had already arrived at the scene. These respondents argued that if it weren't for these online messaging platforms, this issue might have ended up as only a small skirmish on the banks of the lake, and probably would have dissipated after a few slurs were hurled at one another; the intensity of the intimidation attempt was aggravated by means of quick mobilisation through the phone. Similarly, other respondents in this cohort showed on their devices, well-crafted and artistically drawn posters circulated on social media by right-wing political activists about how the caste system ensured the pristine and divinely ordained social order where everyone knew their place and, for the collective good, had better stay there.

> if anyone makes a casteist statement in public, we can all gather around that person and defeat him ideologically and using argument. Maybe even slap a legal suit. However, can we hunt each of these messages and posters and delete them? By the time we see one, a thousand people would have received it, and maybe even get convinced slowly but surely! They even call for bringing back the traditional order of 'keeping people where they ought to be', for 'social stability' and 'traditional values.' This is the new menace attempting to thwart our movement.

The phone, these respondents reasoned, had provided a new window and had magnified the momentum for this sort of counter-activism to subdue their decades-long toil around the Dalit cause.

Disentangling the convergence

The previously described findings, in themselves, may not be novel. They may well be applicable to other social groups in other locations too, not just Dalits in peri-urban Bangalore. For instance, Rashmi (2017) has demonstrated clearly how other segments of the underclass such as security guards, domestic workers, housekeeping staff, janitors, transport staff, and others experience the mobile phone in rather similar ways.[3] The issue here is to understand that such outcomes, even if not distinctive, are not freestanding or autonomic, and do not arise devoid of the frames they operate within. These socio-technological outcomes are actually the result of a convergence between actualities that may be independent of one another. These realities, rooted in the sociological and political ambience of the locale, are what decree the technological experience of a social group. In the context of our enquiry, these elements in association with one another are what decide whether ICTs actually do assure inclusive economic participation for disadvantaged social groups. To reiterate the question at hand: have new digital communication technologies been insufficiently harnessed by historically disadvantaged social groups in this region, or have they reinforced caste-based socioeconomic exclusion? The answer, unfortunately in the affirmative, is the consequence of three elements – (a) the durability of caste in peri-urban metropolitan India, (b) the social construction of the usage of ICTs (in this study, mobile phones), and (c) the myopia in the conventional policy and popular understanding of the digital divide in India.

The durability of caste in metropolitan India

One macroeconomic episode that repeatedly reared its head while interviewing members of Cohort B was the process of economic liberalisation that the Indian economy underwent from the 1990s onward. Many respondents within this cohort agreed that it was economic liberalisation, first and foremost, that gave the commencing liftoff to their rise. However, a close look at the evidence in the literature clearly shows how liberalisation benefited mostly those who were already on a rise but not necessarily everyone who started from a setting such as those of Cohort A. In other words, economic liberalisation came with tremendous possibilities of upward mobility, which millions of individuals in India harnessed, but it did not *assure* any social group of an economic ascent. Caste remains – even today, a quarter century after industrial reforms were established – a prime source of social and economic capital. These reforms strengthened mostly the already-capitalised social groups, while being indifferent to the social exclusion of those lacking this critical stock of social and economic capital to catch the tide of economic growth and prosperity (Chalam, 2011). Dalits in urban India, who comprise the large majority of these economically excluded social groups, still for the most part do not have the requisite access to beneficial networks and markets, and are left to operate in informal, casual markets even within metropolitan areas (Das, 2010). They generally do not possess the means and opportunities to overcome vulnerabilities or economic exclusion unless they are politically connected, or happen to be fortunate enough to capture ever-elusive opportunities in potential small-business sectors (Duncombe, 2006) – in the manner that Cohort B achieved. Economic outcomes thus clearly differ between Dalits and most other social groups in the urban or peri-urban economy, as caste is selectively reworked in these settings to create unfavourable terms of exchange for Dalits, excluding them from critical information and credit sources (Harriss-White, 2004; Harriss-White et al., 2014). A large proportion of emerging businesses in peri-urban India operate on intra-caste or intra-class insular interactions between individuals and groups, especially when they have access to political power in the region. This works in the converse for Dalits, who remain entrenched in their networks that are already in penury. Prakash (2015a) has elaborated extensively on the stifling of economic upward mobility among Dalits, especially those who wish to engage in entrepreneurial ventures; he provides a strong argument on how perversions in state, civil society, and market at their intersection have brought about rigidity in credit access, network information access, and other resources for Dalits in urban India. Conditions such as these have immediate footprints on the kind of business or financial network contacts that Dalits possess. Dalits in Cohort A, operating at the margins, to be reconciled to the thought that their phones will never hold contact details of the wealthy and powerful. A general sense of disappointment prevails in communities such as Cohort A, in which people can clearly see but not necessarily articulate, the fact that the mobile phone revolution has bypassed them due to the fact that caste is still sturdy and resilient in peri-urban Bangalore.

However, individuals in Cohort B appear to have escaped these conditions, and have succeeded in latching on to the wave of entrepreneurship in metropolitan Bangalore that emerged from the late 1990s onward. This wave emerged from the expansion of the unorganised sector, which dominates the metropolitan economy outside of the organised or caste-led sectors of the city (Kapur et al., 2014). Respondents from Cohort B recalled another factor that permitted their rise: small businesses that have become a defining characteristic of urban economies all over India over the last two decades, a trend is reflected also in the literature on Dalits and the economy (Verma, 2015). Whether these opportunities arose as spin-off effects of the formal sector expanding in Bangalore (such as manufacture of spare parts or setting up of small ancillary units), or whether it is small businesses (such as utilities and household services) operating within neighbourhoods, even a small bit of capital was sufficient to latch on to these openings that spawned in the 1990s and 2000s. Several individuals in Cohort B reported to have started small welding units in the 1990s, which then slowly flourished into small units engaged in iron works, to full-fledged shops and retail outlets manufacturing and selling commodities such as gates for households or rolling shutters for shops and garages. Another entrepreneur related how he began with selling "export-reject" shirts (apparel rejected by importers due to tiny flaws in their manufacturing) on the street, slowly built contacts with bigger sellers, and after 20 years now is a major exporter. All this success, first, because they migrated into the city and seized the smallest opportunities available in the informal economy, and second, because they had opportunities to connect with others – even if not with upper castes, with more well-to-do individuals in the middle and lower castes who were not Dalit. Crucially, this sense of connectedness with other members of the underclass – who could be Muslim, or those who belong to Other Backward Classes (OBC), or even new migrants from Western India – appears to have prevailed. Within the city, these individuals in Cohort B would have in all likelihood recognised that networking with people across a spectrum of castes is an entry point to economic prosperity. Some respondents stated that they managed to "overcome caste" by connecting and merging their caste networks with those of other such subaltern groups in the city, creating a win-win collaboration across a palette of castes. Solidarity with similar subaltern groups and networks was the way out, and the way up.

Hence, the durability of caste in peri-urban and metropolitan Bangalore is a lived reality. The drastically different operationalisation of this reality for members of Cohorts A and B has led to contrasting socio-technical outcomes between them.

The interpretation of the digital divide

Wider patterns of inequity in urban India along the lines of caste are mirrored in inequalities with ICTs, which is not solvable by the simple ownership of a device by one and all (Saith, 2008). Owning or not owning a phone cannot remain the only condition that characterises the digital divide in peri-urban or metropolitan India. The trench of the digital divide is not crossed simply by purchasing a device and beginning to use it to place phone calls; if this were so, the digital divide as is conventionally understood can effortlessly be bridged by handing out devices, maybe even free of charge, to the "have-nots." This binary mode of thinking oversimplifies people and social groups as simply digital haves or have-nots in an archaic and trivialised sense (Qiu, 2009). If the digital divide were truly binary in this sense, the achievability of "universal connectedness" in an economy would be a task that could probably be successful in under a year.

To expand our understanding of the forms that a digital divide can actually take, Qiu (2009) has revealed through his monumental study in urban and peri-urban China that there is a class of ICT users who can be categorised as the *information have-less*. These are a vast group of people who include migrant informal workers, those who seek to escape agrarian distress, those who have highly precarious and vulnerable working lives, and those who generally constitute the underclass in the city and the peri-urban. There is a strong likelihood that every one of these individuals possesses a mobile phone, or are even able to search something on Google, to select videos on YouTube, or even send an e-mail through an independent account. Have these truly crossed towards the favourable side of the digital divide canyon? Even on preliminary thought, and certainly on empirical observation of this category of individuals, the answer may not be a simple affirmative. In the case of India, this calls to mind the Dalit neighbourhoods and communities that are either dispossessed and forced to leave urban areas or are forced out of villages into the peri-urban, settling in habitats such as seen in Cohort A. They may possess a device such as a smartphone, will in all likelihood be able to select a video of their favourite song on YouTube, possibly download a film that the family can watch, share content, and so on but tangibly lack the effective participation in the digital revolution either as creators of technological products and services or as customers who are beneficial for information feedback to the producers; rather, they are simply passive consumers of digital technologies on the margins, participating in the digital experience purely for casual interpersonal communication or entertainment.

Conceptualising the peri-urban subaltern in this manner opens up a fresh and far more realistic understanding of the digital divide in its more realistic form, which, as Qiu (2009) proposes, builds on the existing digital divide and takes it beyond a superficial impression. There is hence a populous category in the digital divide spectrum that the Indian policy discourse on ICTs has entirely overlooked. Apart from the obvious recognition that those among the poor who possess digital devices must be taught to *use* them better, this is often stated almost as a pedagogic prescription and is very restrictive in terms of digital participation (as stated earlier, they remain passive consumers). Programs such as Digital India or the National Digital Literacy Mission speak of training lower-rung government workers and other social workers in ICT skills to access e-governance services, or of improving digital infrastructure in terms of device ownership and broadband highways. Even those objectives, of these programs that directly address empowerment, dwell on information access through mobile phone internet, with little interest in whether or not these spaces in the digital sphere will ever be accessed by users. These programs are noble endeavours indeed, with little doubt regarding their intentions to expand the digital experience for the demographically gigantic underclasses in India, but they are certainly myopic of social realities of these economic classes and subaltern castes.

The information have-less, who constitute almost the entire Cohort A and some members of Cohort C of this study, may have information (say, about state programs or information on subsidies) but are rarely *informed* users of digital technology (Webster, 2006). They may own technological devices, and may use them very recurrently through the day, but they are still alienated and disempowered in a larger sense of citizenship and digital participation, which is clearly because these technologies operate within existing social conditions without changing them (Qiu, 2009). This is similar to the heavy critique throughout the literature on the separation of social inequity from digital access, arguing for a conceptual deepening of "digital social inequality" (Halford and Savage, 2010; Lupton, 2015). Interestingly, there may even be a digital divide inside the home, as domestic spaces are mirrored in the use of digital devices, creating another digital divide (Bell, 2006a, 2006b); in fact, this is very evident in our findings, as seen in

107

Table 5.1. Once we open up our understanding of digital inequities as not simply a matter of ownership of device, the mammoth scale of the task and agenda ahead for reorienting technology policy becomes evident (Halford and Savage, 2010). Naiveté in the state policy discourse around ICTs results in the interpretation of the digital divide as simply e-inclusion in the sense of ownership of devices and access to information, rather than effective participation, demonstrating the flawed inclination of this policy orientation towards technological determinism and solutionism (Fortunati, 2008; Morozov, 2011, 2013).

Understanding the information have-less is enriched by appending concepts from feminist literature on the diversity of users of a technology, which differentiates users into "end users," "lay end users," and "implicated actors" (Oudshoorn and Pinch, 2003; see also Wajcman, 2004, 2010). While the first two terms refer to those who are affected downstream by innovation and those who have been excluded from expert discourse, "implicated actors" include those who have been either not physically present but discursively constructed and targeted by technology developers, or those who are physically present but who have been generally silenced, ignored, or made invisible by those in power. Extending this typology to our own enquiry, members of Cohort A, too, all own devices but have been excluded in the meta-narrative of the promised prosperity that mobile phones are supposed to bring them.

At a broader level, according to the UNDP's human development data for 2018, there are about 78.8 mobile phone subscriptions for every 100 individuals in India. In this study, almost all individuals in all cohorts possess mobile phones, and in principle have access to general information through the phone. Has the digital divide been bridged for Dalits in peri-urban Bangalore, let alone for India as a whole? If the focus is on simple access, the answer inclines towards a yes. If we broaden this to a spectrum of have, have-not, and information have-less, we arrive at a more realistic understanding of the actual digital divide in peri-urban India, as well as of the current socio-technological circumstances of each of cohort in this study.

The social construction of mobile phone usage

The third component of the convergence that leads to differential socio-technological outcomes concerns the manner in which social groups (whether caste, class, age group or, in this study, cohort) idiosyncratically interpret and attribute meaning to that technology. In this analysis, we apply and extend the social construction of technology (SCOT), as conceptualised by Bijker (for instance, Bijker, 1987, 1997, 2010; Bijker and Pinch, 1987; Bijker et al., 1987).

A multiplicity in meaning about a technological device results in the device experiencing flexibility in interpretation by various social groups engaging with it. The distinct interpretation of this device within a social group arises due to interactions and discussions within its members around the best uses for, or the ideal operation of, that device. In this process, over time, there emerges a certain stability in meaning and interpretation, resulting in an interpretive stabilisation within that social group about that device; though this stabilisation is seldom conceived to be permanent within the life cycle of the device, the timeframe within which the interpretation of the device "settles" is broad enough for the interpretation to capture the psyche of most of the social group's members, especially if they are concentrated around a limited socio-geographic locale. All these processes operate within (and in turn, build up) what is termed as a "technological frame," which is a paradigm incorporating the tacit knowledge, procedures, goals, and techniques of technological problem solving around that device.[4] In this manner, interactions between members of a social group play a significant role in the evolution of a particular form of technology and its interpretation for that group.

In this study, however, it can be seen that the cohorts do not play a role in the technological evolution of mobile phones. In fact, the contribution of our cohorts even as users of the technology might not have shaped its creation.[5] Rather, our cohorts have shaped the interpretative flexibility (within themselves) in how these mobile phones are *used* in interpersonal interactions. Cohort members interact, attribute meaning, and reach a consensus around the "ideal" or "best" usage of a device. Each cohort interprets the device in its own manner, while groups within each cohort (for instance, young men, the middle aged, or women) share an interpretation of what a device means to them and what a mobile phone is useful for, among other such concerns around the device. Over time, these interpretations stabilise and end up socially constructing not the phone itself but its *usage* in that social context. Supplementing this perspective is an understanding of how each cohort has "domesticated" a mobile phone. It would be worthwhile to spend some time familiarising ourselves with this conceptualisation, to complement the SCOT framework we have just employed.

According to the domestication framework (Silverstone et al., 1992; Silverstone and Hirsch, 1992; Silverstone and Haddon, 1996), there are four phases to the domestication of a technology – appropriation, objectification, incorporation, and conversion. Appropriation of a device occurs when it is sold to a consumer and owned by them, while objectification processes reveal the norms and principles of the individual's or group's sense of itself and its place in the world. Appropriation and objectification occur in (continually) constructed spatial environments, never in isolation. The focus for the analysis here, however, is on the third and fourth phases – incorporation and conversion. While the phase of incorporation is where the device is used in the routines of daily life, the phase of conversion is where the device shapes relationships between its users and other people. Incorporation, as can easily be understood, is heavily determined by social variables such as gender and age.[6] Similarly, conversion defines the relationship between the user and the external world (in this case, outside the household, caste group, or neighbourhood), in the process claiming a status in the wider society. Hence, the complexity of the experience of a device such as a mobile phone intensifies, as it is not simply a technological *artefact* restricted within the individual domain (or even within a household, such as in the case of a refrigerator) but rather a *medium* of interaction between individuals and social groups, therefore doubly articulating both private and public domains (Silverstone et al., 1992). In the case of this study, the social construction of the usage of the phone is conditioned greatly not only by ownership (and hence prestige), or the relationships between individuals and their peers but importantly by incorporation and conversion, which are significantly based on the contemporary subaltern condition or slow upward socioeconomic mobility. Interactions around an artefact, within (and sometimes between) social groups associated with it, results in the stabilisation in meaning and of the general technological frame around the device for each social group (Bijker et al., 1987). This is why members of each cohort in this study have contrasting understandings about their experiences with mobile phones.

These differing experiences within the same caste group, or even within cohorts, appear to ignore the "script" attributed to this digital device by designers and policymakers, who subscribe to its alleged transformative powers. Technological devices are usually defined within a framework of action, taking into account the actors and spaces within which they are scripted to operate. In other words, technologists anticipate the skills, motives, and behaviour of the represented users of this device, attributing to them certain assumed competencies, actions, and responsibilities, all of which then become materialised in the design of the device (Oudshoorn and Pinch, 2003). However, this idealised attribution – echoing an attitude of technological determinism – may not fructify. Akrich (1997) argues against restricting a device's attributes to designers (and, in this case, policymakers). One must appreciate and take stock of the form and meaning constituted by, and constituted of, the technological device, to understand the *de*-scription of the device that results from the mismatch between the users as imagined by the designers (and policymakers), and actual usage of the phone.[7]

The experience of a technological device thus rests significantly on how it constrains users in the ways that they relate to it as well as to one another, and in addition, to how they reshape the usage of the device with these interactions. In designing an artefact such as a mobile phone, the user is also "constructed" in the process as one who completes both the function and vision embodied in it, and hence the entity that needs to be harnessed to reinforce the place of the device in contemporary socioeconomic life (Silverstone and Haddon, 1996). Reality is far removed from these idealised designs and assumptions, which is the reason why a scripted approach cannot explain the variety of technological outcomes and experiences, such as among our cohorts, or for that matter by Dalits at large as compared to other, more economically successful castes. While scripts may foretell prosperity, the reality of the socio-technical outcome, due to the social construction of the usage of a device, can result in increasing or decreasing returns. It is uncertain as to which competing technologies will emerge dominant in any given situation; it may even materialise that the second-best gains dominance due to rapid lock-in, which can emerge even on account of small historical events (Arthur, 1989; Cowan, 1991). This can well apply to the usage of digital technologies also. However convincing the intention may be, and however one might remain predisposed towards an all-positive usage of ICTs, unexpected events on the ground, expected realities and recognised historical circumstances, or any other such conditions exogenous to the technological device itself, can solidify the usage of the technology within a social group by the group's own interpretation and construction of usage.

Conclusion

This study illustrates how a technology – even its *usage* – is a socio-material product, combining artefacts, people, cultural meanings, and knowledge (Wajcman, 2004). Mobile phone experience among members of deprived castes in peri-urban Bangalore is hence not only a function of their immediate knowledge of using these phones but also draws from the dynamics of peri-urban work and life, the durability of caste in the peri-urban setting, in what form a digital divide manifests itself, and the role of ICT in contemporary Indian peri-urban society. There is nothing inevitable about the way a technological device ought to operate within a society or a social group; technologies may be deployed highly heterogeneously and have differentiated outcomes across society.

The peri-urban, therefore, much like a city itself, is a socio-technological outcome in which a device and social groups dealing with that device interact through the political, cultural, and economic dramas of daily life (Guy and Karvonen, 2011). The presence of technology and the closing of the digital divide as commonly or conventionally understood are clearly insufficient, as technology itself cannot budge social inertia and may even amplify social conditions (Toyama, 2015). A technology can amplify the inclinations of individuals and groups, such as those who wish to pursue merely entertainment (such as young men in Cohort A), or for those who wish to pursue entrepreneurial ventures (like members of Cohort B), or for those who wish to kindle dissent and cultivate social movements (Cohort C). Because of this, low-cost technology need not always fight inequality as they are scripted to do so, as technologies often act as reflections and amplifications of social inequities instead of remedies; these technologies serve not as bridges but as jacks (Toyama, 2015: 49).

Our understanding of the role of technology in alleviating historical deprivation must be significantly deepened, as there is clearly a huge overlap between the virtual and the real (Hammersley and Atkinson, 2007). Mobile telephones might have only multiplied efficiency and speed of communication, but the larger outcome of whether or not ICTs such as mobile phones have been insufficiently harnessed by historically disadvantaged social groups to overcome exclusion, and thereby combat deprivation over the long term, is contingent on the persistence of social structures, the understanding of what a digital divide really is, and how social groups interpret specific technologies (Castells et al.,

2007; Toyama, 2015). By itself, technology cannot do much in the direction of social change. Simply making it all-pervasive in life and work does not create a knowledge society or a smart city. By themselves(and even in an urban or peri-urban setting, with all its notions of liberation from caste and identity), ICTs do not automatically radicalise or politicise people who have been historically deprived (Morozov, 2011, 2013); this has clearly been the case with Cohort A. Socio-technological outcomes depend greatly on the convergence of a series of factors. This is also why peri-urban Bangalore has emerged as an archetypical showcase of socioeconomic inequity, hence digital inequality as well. Even for the formal sector of the ITES industry in Bangalore, there has been a recognition that a large proportion of people, even in the immediate locale of Electronics City, have been excluded – which has created a local but highly significant and expanding urban dualism (Upadhya and Vasavi, 2006; Upadhya, 2007). This urban dualism easily expands to digital dualism, much beyond the sense of a conventional "have" or "have-not" digital divide.

All this has very deep implications, especially for Dalits. According to the Census of India 2011 and the Socio-Economic and Caste Census of India 2011, the literacy rate among Dalits is approximately 60 percent, compared to an over-all national rate of 74 percent. Dalits also have low average monthly income, dwindling participation in secure employment, and very marginal land owner-ship rates. Given this reality, a mobile phone ends up as nothing more than a social communication device to maintain informal contact with friends or extended family for many, or, for the younger individuals, as an entertainment device. More positive outcomes for Dalits will be possible with greater political participation, increased access to quality education and healthcare, and other fundamental socioeconomic variables; not just increased ICT ownership and internet access, as what state programs wish to do (see also Kamath, 2017). Technological trajectories can empower the marginalised only when the frame-works of equity and justice are in place, without which connectivity will only aggravate exclusion (IT for Change, 2015). As Toyama (2015: 54) has noted, "technology results in positive outcomes only where positive, capable human forces are already in place."

This returns me to the question I ask at the beginning of this study in Chapter 4 – have mobile phone technologies been insufficiently harnessed by historically disadvantaged social groups, or have they assisted in the reinforcement of these exclusions for some? The evidence unfortunately points towards the affir-mative. For the most part, Dalits do not entirely remain passive onlookers in the technological trajectory of the mobile phone in India, as seen in the experi-ences of Cohort A, but they are not active agents either. They are generally what I would term as "passive participants" in this story, who are subjected to a largely patronising attitude of state and market with respect to enriching their technological experiences. The intentions behind state programs arise from well-accepted notions that ICTs facilitate greater participation in the development process, especially in metropolitan regions such as Bangalore. Given that ICTs have pervaded life and work in urban India over the last 15 years, this optimistic outcome, especially for the urban poor and historically deprived, ought to have materialised. But access is rarely translated into actual and automatic opportunity to drive upward socioeconomic mobility (Kamath, 2017). Also, given that there is a vast population of the information have-less in peri-urban areas, these optimistic outcomes may only play out on paper. All this, despite the fact that mobile phone penetration is nearly 80 percent in India, with more than a billion devices currently in use.

Notes

1 An earlier and much shorter version of this chapter has appeared as A. Kamath (2018), "'Untouchable' Cellphones? Old Caste Exclusions and New Digital Divides in Peri-Urban Bangalore," *Critical Asian Studies*, 50(3): 375–394. Used with permission of Taylor & Francis Ltd, www.tandfonline.com, DOI: 10.1080/14672715.2018.1479192.
2 For a very recent instance of collective political organisation around the Dalit cause in northern and central India, see Daniyal (2018). Platforms such as WhatsApp and Facebook were employed not only to spread messages, poetry, and infographics but also to exhibit solidarity by replicating profile photos (or display pictures) that had strong messages on them.
3 These individuals would be categorised as the "information have-less," a concept this chapter discusses in the following subsection.
4 This is similar to a Kunhian technological paradigm; the difference, however, being that a technological frame structures the interactions among the members of social groups (beyond scientists and engineers) that are associated with this technology (with the possibility of varying degrees of inclusion among individuals within one technological frame, as well as that of an individual participating in multiple technological frames), while a Kuhnian paradigm is intended for understanding scientific communities. That is, this is a frame with respect to the technology in question, and not a technologist's frame (Bijker et al., 1987).
5 Though this is altogether a separate enquiry that will be very valuable to the literature on the role of users in technological development and social construction of a technology.
6 In fact, incorporation can even reinforce the culture of the technology; for instance, its embedded masculinity (Oudshoorn and Pinch, 2003).
7 As Akrich (1997) argues further, designers often tend to accuse the users of having "misused" a device, in the event that the device ends up unsuccessful.

References

Akrich, M. (1997) 'The De-Scription of Technical Objects', in Bijker and Law (eds.) *Shaping Technology/Building Society: Studies in Socio-Technical Change*, The MIT Press, London and Cambridge, MA.

Arthur, W.B. (1989) 'Competing Technologies, Increasing Returns, and Lock-In by Historical Events', *The Economic Journal*, 99(394): 116–131.

Bell, G. (2006a) 'Satu Kelugara, Satu Komputer (One Home, One Computer): Cultural Accounts of ICTs in South and South East Asia', *Design Issues*, 22(2): 35–55.

Bell, G. (2006b) 'The Age of the Thumb: A Cultural Reading of Mobile Technologies from Asia', *Knowledge, Technology, and Policy*, 19(2): 41–57.

Bijker, W.E. (1987) 'The Social Construction of Bakelite: Toward a Theory of Invention', in Bijker, Hughes, and Pinch (eds.) *The Social Construction of Technological Systems*, The MIT Press, Cambridge, MA.

Bijker, W.E. (1997) *Of Bicycles, Bakelites, and Bulbs*, The MIT Press, Cambridge, MA.

Bijker, W.E. (2010) 'How Is Technology Made? That Is the Question!', *Cambridge Journal of Economics*, 34(1): 63–76.

Bijker, W.E., Hughes, T.P., and Pinch, T. (1987) *The Social Construction of Technological Systems*, The MIT Press, Cambridge, MA.

Bijker, W.E., and Pinch, T. (1987) 'The Social Construction of Facts and Artifacts: Or How the Sociology of Science and the Sociology of Technology Might Benefit Each Other', in Bijker, Hughes, and Pinch (eds.) *The Social Construction of Technological Systems*, The MIT Press, Cambridge, MA.

Castells, M., Fernandez-Ardevol, M., Qiu, J.L., and Sey, A. (2007) *Mobile Communication and Society: A Global Perspective*, The MIT Press, Cambridge, MA and London.

Chalam, K.S. (2011) *Economic Reforms and Social Exclusion: Impact of Liberalisation on Marginalised Groups in India*, Sage, New Delhi, Thousand Oaks, London, and Singapore.

Cowan, R. (1991) 'Tortoises and Hares: Choice among Technologies of Known Merit', *The Economic Journal*, 101(407): 801–814.

Daniyal, S. (2018) 'The WhatsApp Wires: How Dalits Organized the Bharat Bandh without a Central Leadership' *Scroll*, accessed 7 April 2018: https://scroll.in/article/874714/the-whatsapp-wires-how-dalits-organised-the-bharat-bandh-without-a-central-leadership

Das, M.B. (2010) 'Minority Status and Labour Market Outcomes: Does India Have Minority Enclaves?', in Thorat and Newman (eds.) *Blocked by Caste: Economic Discrimination in Modern India*, Oxford University Press, New Delhi.

Duncombe, R. (2006) 'Analysing ICT Applications for Poverty Reduction via Micro-Enterprise Using the Livelihoods Framework', Working Paper 27, IDPM, University of Manchester.

Fortunati, L. (2008) 'Re-Thinking E-Inclusion', Prati CIRN 2008 Community Informatics Conference: ICTs for Social Inclusion, 12 July.

Guy, S., and Karvonen, A. (2011) 'Using Sociotechnical Methods: Researching Human-Technological Dynamics in the City', in Mason and Dale (eds.) *Understanding Social Research: Thinking Creatively about Method*, Sage, London.

Halford, S., and Savage, M. (2010) 'Reconceptualising Digital Social Inequality', *Information, Communication, and Society*, 13(7): 937–955.

Hammersley, M., and Atkinson, P. (2007) *Ethnography*, Third Edition, Routledge, London and New York.

Harriss-White, B. (2004) *India Working: Essays on Society and Economy*, Contemporary South Asia, Foundation Books and Cambridge University Press, UK, New Delhi.

Harriss-White, B., Basile, E., Dixit, A., Joddar, P., Prakash, A., and Vidyarthee, K. (2014) *Dalits and Adivasis in India's Business Economy: Three Essays and an Atlas*, Three Essays Collective, India.

IT for Change (2015) *Annual Report 2014–2015*, IT for Change, Bangalore.

Jeffrey, R., and Doron, A. (2013) *Cell Phone Nation*, Hachette India, New Delhi

Kamath, A. (2017) 'In India, Accessible Phones Lead to Inaccessible Opportunities', *The Wire*, 24 November.

Kapur, D., Babu, D.S., and Prasad, C.B. (2014) *Defying the Odds: The Rise of Dalit Entrepreneurs*, Penguin Random House India.

Lupton, D. (2015) *Digital Sociology*, Routledge, Oxon and New York.

Morozov, E. (2011) *The Net Delusion: How Not to Liberate the World*, Penguin, London.

Morozov, E. (2013) *To Save Everything, Click Here: Technology, Solutionism, and the Urge to Fix Problems That Don't Exist*, Penguin, London.

Oudshoorn, N., and Pinch, T. (2003) 'Introduction: How Users and Non-Users Matter', in Pinch and Oudshoorn (eds.) *How Users Matter*, The MIT Press, Cambridge, MA.

Prakash, A. (2015a) *Dalit Capital: State, Markets and Civil Society in Urban India*, Routledge, New Delhi.

Qiu, J.L. (2009) *Working-Class Network Society: Communication Technology and the Information Have-Less in Urban China*, The MIT Press, Cambridge, MA and London.

Rashmi, M. (2017) 'The Digital Others', *Seminar*, 694: 40–43.

Saith, A. (2008) 'ICTs and Poverty Alleviation: Hope or Hype?', in Saith, Vijayabaskar, and Gayathri (eds.) *ICTs and Indian Social Change: Diffusion, Poverty, and Governance*, Sage, New Delhi.

Silverstone, R., and Haddon, L. (1996) 'Design and the Domestication of Information and Communication Technologies: Technical Change and Everyday Life', in Mansell and Silverstone (eds.) *Communication by Design: The Politics of Information and Communication Technologies*, Oxford University Press, Oxford, UK.

Silverstone, R., and Hirsch, E. (1992) *Consuming Technologies*, Routledge, London.

Silverstone, R., Hirsch, E., and Morley, D. (1992) 'Information and Communication Technologies and the Moral Economy of the Household', in Silverstone and Hirsch (eds.) *Consuming Technologies: Media and Information in Domestic Spaces*, Routledge, London and New York.

Toyama, K. (2015) *Geek Heresy: Rescuing Social Change from the Cult of Technology*, Public Affairs, New York.

Upadhya, C. (2007) 'Employment, Exclusion, and Merit in the Indian IT Industry', *Economic and Political Weekly*, 42(20): 1863–1868.

Upadhya, C., and Vasavi, A.R. (2006) 'Work, Culture, and Sociality in the Indian IT Industry: A Sociological Study', Report Submitted to the Indo-Dutch Programme for Alternatives in Development, National Institute of Advanced Study (NIAS), Bangalore.

Verma, V. (2015) *Unequal Worlds: Discrimination and Social Inequality in Modern India*, Oxford University Press, New Delhi.

Wajcman, J. (2004) *Techno Feminism*, Polity Press, Cambridge.

Wajcman, J. (2010) 'Feminist Theories of Technology', *Cambridge Journal of Economics*, 34(1): 143–152.

Webster, F. (2006) *Theories of the Information Society*, Third Edition, Routledge, London and New York.

6

A TECHNOLOGICAL PANACEA FOR WOMEN GARBAGE COLLECTORS

With Neethi P. and Saloni Mundra[1]

In the last section of this book, I present a study, undertaken with two other co-researchers, on a workforce that has rarely been studied in the academic literature – garbage collection workers, or *pourakarmikas*,[2] in Bangalore city. With this study, we focus on gendered experiences of a technological intervention – a biometric attendance system – that is not in itself novel but was introduced in the everyday working lives of these female informal workers for the purpose of streamlining a number of wage-related issues faced by them individually, as well as overthrowing substantial financial fraud affecting the city corporation of Bangalore. While a number of issues for the corporation were ironed out due to the technological intervention, the everyday working lives of these women appear to have changed little. Would this outcome imply that the technology was a "failure" for workers? Or is the picture more complex? In order to understand why new issues have surfaced subsequent to the introduction of the biometric attendance system, we need to dive below the surface and understand the dynamics operating under the technological experience. Upon delving deep enough, and after encountering a few other attributes such as informality and caste, what stands out significantly is gender.

Gender stands as the nucleus of their everyday technological experiences, not simply because of the established fact that technologies are intrinsically embedded with the politics of gender but because these technological experiences in this case are an extrapolation of their everyday *work* experiences, which are very perceptibly entrenched in the problematic gendered nature of informal work. To add to the concoction, there are inseparable issues of caste that operate at the crux of these experiences. In other words, power equations that have long been gendered and caste-driven in the organisational system within this line of work, have neatly been reincarnated in the new technological regime. This may not have been foreseen (and is still feebly appreciated) by the authorities who introduced this device into their work experience in order to abrogate staggering corruption from the part of garbage collection contractors engaged by the city corporation.

116

The body of thought and global empirical evidence on the highly gendered nature of informal work is far from scanty, and this study draws from this well-spring to understand the lived realities of everyday work among women garbage collectors in metropolitan Bangalore. In parallel, the body of literature on the feminist politics of technology is not novel but has rarely been employed to understand technological experiences among a subaltern (in terms of work informality, caste, and gender) yet populous informal workforce within an urban setting in a developing country. It is therefore by bringing together these two conceptual traditions, and deploying gender as the principal analytical perspective, that we propose that the character of everyday technological experiences is a derivative of the everyday work experiences. Therefore, we also address the larger question of how technological experiences are deeply embedded in sociological contexts; in this case, of multiple attributes including caste, gender, and informality.

We draw from empirical material, secondary sources from civil society organisations, and journalistic chronicles that record the travails of the nearly 12,000 pourakarmikas who manually collect and clear the garbage generated by a metropolis of nearly 10 million. Let us begin with understanding the sector that these workers labour within.

Solid waste management in Bangalore city

The Health Department, within which the solid waste management cell, of the greater Bangalore city corporation or *Bruhat Bengaluru Mahanagara Palike* (BBMP) handles solid waste management in the city. Under the Commissioner of the BBMP is the Special Commissioner of Solid Waste Management. Below this level is a series of other officials and staff, which finally reaches down to the level of the Ward (a collection of 6,000–38,000 households), where the Health Inspector is in command. It is under this level that the teams of garbage collectors operate, collecting solid waste (food and non-food, colloquially termed wet and dry) door-to-door, and street waste (which includes a vast variety of discarded items and material, fallen leaves and foliage, and general garbage strewn by people) from the ward assigned to groups of pourakarmikas. According to BBMP (2017b), pourakarmikas are provided a pushcart trolley (one trolley allotted for every 120 households and shops) assisted by an auto-rickshaw of a 500-kilogram capacity for each block (around 750 households). Generators of bulk waste such as hotels and restaurants, and other commercial establishments, have a different system of solid waste collection.

By the mid-1990s, solid waste management for a burgeoning city started becoming unwieldy for the BBMP to handle alone. Convincing itself that the task ought to be outsourced, the corporation, in steadily expanding phases, began to subcontract out this responsibility to a series of contractors, beginning the establishment of a system of informality in its operations below a certain level. Thus, a critical point in this system under the health inspector emerged

Table 6.1 Pourakarmikas and associated workers in solid-waste management (as of July 2017)

	Total	Female	Male
Pourakarmikas	16,505	10,999	5506
Drivers	2836	48	2788
Helpers	2637	203	2434
Supervisors	616	34	582
Total workers	**22,594**	**11,284**	**11,310**

Source: BBMP and <www.opencity.org>

in the form of the *contractor*, who was in charge of recruiting and managing the pourakarmikas and their supervisors (locally addressed as *mestris*). Apart from pourakarmikas, auto-rickshaw drivers and helpers also featured in the pool of garbage collection workers, all under the control of the outsourced contractor. Even today, those who wish to enrol as pourakarmikas are required to contact the contractors, who then screen and recruit the potential workers; this also applies to the auto-rickshaw drivers and helpers. These informal workers stand apart from the 2,500-odd permanent pourakarmika workers (as informed by the health commissioner's office) who are recruited directly by the BBMP, and managed by the corporation itself. The rest, a mammoth population of around 17,000, are all employed and administered under the contract system spread across the city. It must be mentioned here (which will be discussed in later sections) that for many years, the number of pourakarmikas was purported to be more than 30,000, but revelations by both the state and civil society organizations disclosed that a huge proportion of this number – around a staggering 10,000 to 12,000 workers – were "invisible" or "ghost" workers. The number of contract workers attached to each designation or role is presented in Table 6.1.[3]

The pourakarmika workforce

Pourakarmika workers are generally migrants from rural or very remote areas of Karnataka state, from agrarian settings or manual labour backgrounds, in search of casual work in the city. A sizeable number (as elicited from not secondary data but drawing from conversations with the workers and contractors) immigrate to Bangalore city from border towns in Andhra Pradesh and Tamil Nadu. Several women pourakarmika workers have migrated here through marriage as well. In many cases, these women workers become heads of their households here in the city, supporting young children and ageing dependents at home. Some of them juggle this job with other casual or manual jobs (such as construction work, domestic help, or as shop cleaners) to make ends meet, and to save for anticipated expenses such as a daughter's marriage or illnesses. There is a broad range of work experience among pourakarmikas, varying from five years to sometimes more than 25 years in this occupation, depending on their age, which ranges from 20 to nearly 60.

Most pourakarmikas belong to Hindu Dalit castes or other downtrodden castes, which is not a coincidence as these subaltern castes have historically been relegated to cleaning and scavenging work in private and public spaces, even in urban and metropolitan India (Gooptu, 1993; Prashad, 2000), under heavily exploitative and humiliating conditions. Chitageri (2004) has recognised that they predominantly belong to Kannada-speaking Madiga or Kuruba communities, and Tamil-speaking Pariayan or Chattiayan communities. Given their historically deprived socioeconomic backgrounds, and negligible access and opportunity for formal education, they are unable to secure better occupations and hence are at the mercy of work contractors in the city and other unregulated agents who hunt, recruit, and transport vulnerable labour such as this. Many women pourakarmikas manage to acquire this job through referrals by incumbent women pourakarmikas, who refer the names of potential workers to their respective contractors, who then screen and recruit these aspirants. Workers who initially seek housing in an expensive city such as Bangalore are sometimes accommodated in dormitory-like arrangements with four or five workers living in a room, which again is mediated through the contractors. Hence, being at the mercy of the contractor commences right from recruitment and housing allocation, up until the management of everyday work on the streets.

While the supervisors belong to nearly the same castes as the pourakarmikas, the contractors usually belong to economically well-to-do and politically influential but not necessarily "upper" castes. Contractors are usually in monetary and social proximity to those in political power, and also deal with local mafia and henchmen. However, they claim that this is a necessary part of the job as a contractor, given the plethora of risks in solid waste management involving transport, dumping, access to landfills, skirmishes with rival groups, and so on. On speaking to staff within the BBMP as well as a few contractors, it was revealed that the entire process of collection, aggregation, and transport of solid waste is rather complex. There were knotted political and local equations involved, and the maintenance of spaces for landfills or for burning required constant negotiations with various parties ranging from senior officials in the BBMP right down to local mafias. The contractors also claimed that settling local disputes in areas around the issue of solid waste transport and dumping was entirely their responsibility, and hence the inevitability of political muscle, sources and channels of legal and illegal money, connections across the ranks of police and local government, and nexus with many other actors.

One additional issue alarms the informed observer, which is the stark gender differential in the workforce and task distribution in the sector. Let us briefly visit the gender distribution of tasks on the ground, as well as the task distribution of gender, which is explained in Tables 6.2 and 6.3. This data pertains only to genuine, not invisible or ghost, workers.

We can clearly see in Table 6.2 that while at the pourakarmika level the gender ratio is two-thirds female, a stark opposite is observed for the gender distribution

Table 6.2 Gender distribution by task (figures in %)

	Female	Male	Total
Pourakarmikas	66.64	33.35	100
Drivers	1.69	98.30	100
Helpers	7.69	92.30	100
Supervisors	**5.51**	**94.48**	100

Source: Authors' computations based on data from Table 6.1

Table 6.3 Task distribution by gender (figures in %)

	Female	Male	Total
Pourakarmikas	97.47	48.68	72.23
Drivers, helpers, supervisors	2.53	51.32	27.77
Total	**100**	**100**	**100**

Source: Authors' computations based on data from Table 6.1

for non-manual tasks such as driving auto-rickshaws and helping, besides certainly supervising. Table 6.3 shows a different angle, i.e., among women workers in this sector, an astonishing 97.5 percent of women are engaged as pourakarmikas doing menial and defiling tasks, while among male workers this is distributed nearly equally among those doing pourakarmika tasks and those undertaking other, non-pourakarmika tasks such as driving auto-rickshaws and helping. The sector is therefore populated largely by ground-level pourakarmika workers, who for the most part are female. And importantly, whether we look at the gender distribution of task or task distribution among male and female workers, the sector is predominantly unfavourable towards women by associating them with the more menial and foul tasks of handling the solid waste.

Anticipated turbulences

It is natural and expected that a system such as this (involving a sprawling female informal workforce, the state, and an assemblage of private contractors), and in a sector such as garbage collection (involving political power, historical connotations of caste labour, politics in urban spaces, and state-decreed informality in labour), would eventually touch raw nerves among all stakeholders, simmer in its cauldron, and then erupt in a series of worker struggles across years. At this point, it is essential to study each segment of the general turbulence witnessed among contract garbage collection workers in Bangalore across more than a decade.

The central issue among the workers was the delay in the payment of their wages, often for months. The payment system, traditionally, allowed the contractors to be the conduits through which workers' payments would be channelled, due to the fact that the pourakarmikas were recruited and managed by these con-tractors. There were tremendous delays on the part of the BBMP to transfer these payments, often lasting for six months of continuous overdue. Wages, too, were initially a pittance of around Rs.4000 to Rs.5000. This was despite the fact that most of the workers arrived for work regularly each day, and signed in the manual attendance register handled by the contractor. If the workers decided to agitate against the contractors or the BBMP demanding their wages, there was rampant verbal, physical, and sexual abuse meted out to them by these contractors, which often materialised into deeply derogatory action. In one instance in October 2017,[4] when pourakarmikas of the KR Puram area in Bangalore demanded their wages, the contractor liberally abused them with casteist slurs and threatened to undress himself and rape them, with the offer that the workers would get their salaries if they came over and extracted it out of his clothing. When the work-ers opted to take this up with the police, the contractor in question chased the workers with armed gangs, managing to physically attack the workers. This was entirely possible because of a reported deep collusion between the contractor and certain officials of the BBMP, and because of the socioeconomic status of the workers themselves – a central issue we shall take up in later sections. A climactic point that stemmed out of delayed wages was the suicide of a male pourakarmika in June 2018 after being unpaid for seven months.

A related issue was the demand for a status of permanency in work. As described earlier, nearly the entire pourakarmika workforce is contract labour, recruited and managed by contractors outside of the BBMP. Permanent workers are pre-dominantly male, who wear designated khaki uniforms and enjoy better terms of employment. They are rarely seen out on the streets and instead manage minor tasks, minor civil works, and janitorial services in the various ward offices of the BBMP. The plight of contract workers is manifold. Even those contract workers who had spent decades in this occupation remained as contract workers, with some of them lamenting that they had entered this occupation as newly married girls, and remained contract labourers even after they had become grandmoth-ers.[5] It is not uncommon to meet pourakarmikas who have spent 20 or even 25 years in this occupation with negligible improvement in the nature of the work engagement or work conditions. It was also reported how employees state insur-ance and provident fund benefits did not reach the bulk of pourakarmikas, even though there were ongoing demands for this.

However, the most scandalous issue that plagued the entire system, affecting both the pourakarmikas as well as the BBMP, was the presence of "ghost" or "invis-ible" workers in the payrolls. The contractors to whom the BBMP entirely out-sourced worker management and payment had fabricated the existence of more than 10,000 additional workers on their payrolls, marking their attendance in the manual register every day, and embezzling taxpayers' money from the BBMP.[6] The full wages of genuine pourakarmikas were also compromised to facilitate this ghost workforce, with contractors paying the genuine workers only a part of their entitled wages (in cash) and swallowing the rest, while intimidating these genuine workers into attesting on paper that they received full wages. When pourakarmikas demanded their bank passbooks and ATM cards, they were threatened to be dismissed from service by the contractors.[7] The scale of this outrageous practice was so astounding that a BBMP audit revealed only about 17,000 genuine workers, while payment was being transferred to the contractors for 32,000 workers. With the movement to the new biometric attendance system, the BBMP is said to have already saved nearly Rs.250 crores (around US$35 million)![8]

Struggles over the last decade and more, continuing even today, were not without the creation and support of collective agency organizations and social movements among the pourakarmikas. These movements and unions were not toothless, and were also not deterred despite the verbal and sexual abuse meted out on the workers. Three unions working for the welfare of the pourakarmikas – principally the *Guttige Pourakarmika Sangha*, the *Bengaluru Pourakarmika and Contract-Basis Pourakarmika Association*, and the All India Central Council of Trade Unions (AICCTU) – spearheaded the campaign to address all the previously described turbulences. The *Guttige Pourakarima Sangha*, which stands at the frontline, is a registered trade union associated with the AICCTU, representing contract (hence the Kannada language word *guttige* meaning "contract") pourakarmikas, working for the welfare of these workers since 1996 – especially in the realms of illegality in employment conditions, world regulations, and statutory minimum wage. It was their struggle that bore fruit in 2001, establishing minimum wages for pourakarmikas (Sangama, 2002). With strong support from other civil society organizations in the city (such as *Janashakti*, the Alternative Law Forum, Manthan Law, and the *Safai Karmachari Kavalu Samiti*), several strikes were organized across 2017 where pourakarmikas simply stopped collecting garbage from households and the streets for weeks, choking the city in its own generated filth, and creating public awareness using this stark but effective strategy. The responses of the BBMP and Government of Karnataka that followed their agitation, detailed as follows, was hailed as a "partial victory" by the unions.

As a result of the persistence of the unions and supporting organizations, the BBMP was compelled to take significant corrective steps. One, a wage increase from Rs.5000 to Rs.17,000 (divided into a wage of Rs.14,400 and risk allowance[9] of Rs.3000), was put into immediate effect in June 2017. Another measure involved the provision of direct transfer of wages to the bank accounts of individual pourakarmika workers from January 2018, as opposed to channelling payments through contractors to be disbursed to the workers in cash. Other steps included a promise to regularise and shift contract workers into permanent status in phases, and to include auto-rickshaw drivers and helpers as "pourakarmikas" according to the Karnataka Municipal Corporations Act 1976.

However, the most consequential measure, and the principal subject of our study here, was a technological intervention that facilitated all the measures named earlier – the introduction of a biometric attendance system that was to be deployed from December 2017 onward. This system would replace a practice of hand-marking attendance in a register, as part of a larger overhaul of the payment channels whereby the BBMP would transfer payments directly to the bank accounts of the pourakarmikas based on their attendance, which was in principle a welcome move away from transferring wages to contractors in bulk, and expecting that they would disburse the respective payments to individual pourakarmikas. Attendance calculated on the basis of data would automatically be registered in a central payment system, by which transfer of wages would ideally be tamper-proof and transparent.

Unfortunately, this technological intervention, while indeed succeeding in removing the invisible workers for the most part, has brewed new troubles over the existing ones, and in spirit has not changed the working lives of the pourakarmika workers. While work informality and its associated vulnerability have always been a greater challenge for women workers as compared to male workers across this sector, these issues escalated for the pourakarmika workers by virtue of the socioeconomic intersections at which they live and work. The additional new feature of their operation – the biometric attendance system – has neatly reproduced these structured and often glaring gender relations.

Method

Though a roster of listed pourakarmikas is available on the BBMP website as public information, given the enormous size of the population of pourakarmikas and associated workers, it was impossible to try and undertake a random sample of workers across Bangalore to speak to. Through visits to BBMP offices in south Bangalore and a glance through the regulations on solid waste management, initial information on the contract agencies for south Bangalore was elicited. Discussions with an assistant executive engineer facilitated a lead into Koramangala (Ward 151). Within this ward, blocks 5 and 6 (comprising of around 350 households) were chosen randomly. A list of workers was clearly put up at the Koramangala ward office, along with copies of their identity cards, as well as their designations and photographs. A map of this ward was also provided for assistance.

The initial empirical inquiry was entirely devoted to observation of workers on the streets every morning. Fieldwork was undertaken primarily by one researcher; being female, it was relatively easier to follow these women workers for a while without raising suspicion, and to make initial contact and begin to build rapport with them with a lesser probability of hostility. A few weeks were spent in following them around their work on the streets in blocks 5 and 6, and observing not only their daily routines and struggles but also, importantly, the process of biometric attendance marking every day at either location, at separate times (6 a.m., 10 a.m., and 2 p.m.). Interpersonal relations between workers was also observed carefully. From the teams of pourakarmikas working within these blocks, open-ended interviews for data collection were employed, eliciting a variety of information ranging from their work conditions, recruitment, and relationships of hierarchy; to issues of gender, health, and social capital; to experiences with the new technology. Often, focus group discussions were also undertaken when possible. Apart from workers on the ground and their supervisors, interviews were also held with municipality officials, the neighbourhood association heads. Secondary sources, such as manuals and material on the functioning of solid waste management supplied by the BBMP, were also studied. Importantly, we also combed through material in newspapers and other popular press sources, referred to reports from various civil society organisations (primarily the Alternative Law Forum and Manthan Law), and conducted lengthy discussions with members of those organisations who were directly involved with the pourakarmika struggle.

Everyday experiences around the technology

It was hoped that with the new biometric attendance system, leakage of payments at the contractor level would be quashed, exploitative direct involvement of contractors would be curtailed, phantom workers would disappear from the workforce and payroll, contract workers would transition into permanency, and that work conditions in general would be streamlined. The biometric attendance system, rampant across the organised sector and parts of the unorganised sector in India, made its entry in this most lowly field of work too, with all the previously mentioned ambitious solutionist intentions. However, the reality has only partially played out in this direction, and what has emerged is a textbook case of overlooking social realities while formulating technological interventions. The new biometric system simply, and neatly, dovetailed the existing gendered outcomes of informality in this field of work. Let us see how.

In the process of transitioning from the manual signature-based attendance system to the biometric system, the first task for the BBMP was to audit for the number of genuine pourakarmikas, which was a noble endeavour by itself but which also involved weeding out those individuals who did not classify as "genuine." The official explanation was that the ratio of pourakarmika to city population ought to stand at 1:700 (a decision taken by the BBMP with little external or public consultation) – despite the recommendations of the IPD Salappa Report 1976[10] which advised a ratio of 1:500, which would permit the recruitment of 25,000 more pourakarmikas to cater to the quadrupling of garbage in Bangalore since 1976. On the ground, however, this translated into a rash routine of simply dismissing those workers who were either engaged for less than a year, or those who were over the age of 60. The older workers were simply perplexed as to the future course of action, as they were given only a verbal notice of termination, which was logical because their appointment was also not a formal, documented one. They had no semblance of a payslip, as payments were routinely by cash in hand. They had no valid identification to demonstrate their actual age, and naturally nothing to demonstrate how long they served as pourakarmikas. Even if they were over 60, they were asked to simply leave with no social security or any similar goodwill payment for probably decades of service. This was the very first fallout of the transition towards the biometric system – the reckless termination of hundreds of workers simply because they had no identification to affirm their status and get included into the new technological regime. These anecdotes are, in fact, variants of experiences around other technology-based solutionist initiatives in India, such as the Unique Identification (Aadhar), that have had incidents of wildly excluding individuals from essential state services such as public distribution of food, simply because these individuals could not authenticate themselves to receive an Aadhar card. Hence, even prior to a technological intervention beginning its task, possibilities of impact on life and livelihood abound.

The biometric devices that record the thumb imprints of the pourakarmikas to mark their attendance are placed at spots within neighbourhoods that are supposed to be within easy access to the pourakarmika workers, and are, by regulation, to be handled by the health inspector of the ward and not the contractor. Attendance is recorded three times a day – at 6 a.m. on commencement of work, at 10 a.m. to record half-days (Wednesdays and Sundays), and at 2 p.m. to record complete days at work. The attendance portal is opened for a given period of time, and once closed by the authority handling the device, cannot be reopened for workers arriving late at the attendance spot. This feature intended to ensure that there was no perversion of the system of timekeeping. In the system prior, where signatures were manually recorded, it was easily possible for workers who were more favourable in the eyes of the contractor or health inspector to arrive casually late in the morning or to leave early, fraudulently marking a complete day at work. The biometric system was intended to overcome this issue, which in principle has been successful.

But while genuine and diligent pourakarmikas were envisioned as the final beneficiaries of this new technology, a Pandora's box of new issues has been thrown open. One anxiety that has arisen among the pourakarmikas, in connection with the timekeeping system, is to hasten work or abandon whatever task at hand, to scurry towards the biometric attendance spot so as to not avoid a day's or half a day's leave, as the attendance portal is opened for about 15 minutes at maximum, and cannot be reopened subsequently. Even a two-minute delay after closure at 10 a.m. or 2 p.m. results in a half-day leave, while a small delay at 6 a.m. results in the system registering the worker as not having shown up for the day, which results in a loss of pay. This not only compromises quality of work conscientiously within the workers' own minds but also frequently lands them in trouble with residents of neighbourhoods for hastily done work. Neighbourhood residents have sometimes even *filmed* pourakarmikas dumping whatever unfinished work in hand and closing work for the day, uploading these videos on social media to invite comments on the "shoddy job," without realising that these workers are actually trying to hasten to the biometric attendance spot to avoid a wage cut. In fact, clandestine video recording of especially women pourakarmikas is not rare, and circulation of these videos across neighbourhood associations and even to the BBMP is routine, given that the BBMP keeps its portals open to uploading these clips for the purpose of ensuring customer service. While the biometric attendance system may not have directly caused this serious violation of individual privacy, the introduction of the system appears to have overlooked the fact that these malicious practices are unchecked among the city's citizens.

When there are delays in attendance marking, pourakarmika workers usually miss buses or other transport to return home, where further domestic and other responsibilities awaiting them are also in turn delayed. Hence, in order to rush towards the attendance marking spot, nearly 30 to 40 minutes of work on the streets each day is compromised. While occasionally it is possible to hitch a ride with the auto-rickshaw drivers in their garbage-collecting auto-rickshaws, this may not always materialise, not because of any restrictions from the part of these drivers but simply for the reason that the auto-rickshaws may be collecting garbage at a location far from where a pourakarmika is working at the moment. Male workers suffer from this issue less (as they have transport), while women workers, working on the ground, have to get to the biometric attendance marking spot is mostly by foot, however far they are from it. This gets exacerbated when workers are placed at a great distance from the attendance marking spot, or when they are randomly shuffled across streets that may not be in proximity to the biometric attendance marking spot. Clearly, this differentiation of working conditions between male and female workers (which will be illustrated in greater detail in later sections) has become aggravated following the new technological system.

Connectivity issues with the device are also regular, which delays work at 10 a.m. or departure back home at 2 p.m. Because the device registers only the fingerprint of the worker in person at the designated time, it is impossible to take attendance at a later point in time in the event that connectivity is low or intermittent. Losses in internet connectivity are highly rampant, as often as twice or thrice a week. Occasionally, if the biometric attendance point is at a location that has a weak GPS signal, this can also result in an attendance routine failure. There is a back-up register maintained in the event of a device failure, but in most cases our respondents reported that the individual in authority usually insists on relying on the device alone and awaiting its resurrection whenever it loses connectivity. With male workers, things are far more lenient as they are not made to wait for connectivity, and back-up registers are utilised. The final fallout of an issue in connectivity is the pourakarmikas end up losing a day's or half-day's attendance, hence wages, even if they have worked the entire day. Pourakarmika unions have included these issues in their pressing demands and demonstrations outside the BBMP office. Their wages, while in the earlier system being siphoned off by contractors, are now being compromised due to technical errors – due to which most workers are unable to take home the entitled wage of about Rs.14000 and end up with around half or two-thirds as much, despite working most days of the month.

Another issue that arises is that fingerprints sometimes do not get registered, not because of any malfunction in the device but simply because of grimy or soiled hands given the nature of the job (and in almost all cases, also by virtue of working without gloves). Once again, the female workers are at the losing end, as it is their hands that are grimier than the hands of male workers (again, which we will see in later sections). They are held responsible for fingerprints either not matching or not being registered out of dirty hands, and again are subject to verbal reprimand or wage loss out of non-registering of attendance.

Finally, another issue that is blatant is that the operation of the device, by regulation to be handled by a BBMP authority, the health inspector, is routinely handled by the supervisor or contractor. This is a clear flouting of regulations, which outlines that an individual in the capacity of a state authority and not the outsourced contractor is in charge of attendance, as it is linked directly to wages. By handing this crucial operation back to the hands of the contractor, the system regresses to these women workers returning to the mercy of the contractor's management. Negotiation for women workers is unthinkable given the nature of the reprimand they anticipate. A gendered power structure intrinsic to operations at this level pits women workers at a disadvantage even under the new technology.

An interesting element that was visually observed during fieldwork (which was mentioned in passing but not significantly reported by the women workers, probably on account of internalisation of this discrimination) was greater discipline expected of women pourakarmikas at the attendance marking spot compared to male pourakarmikas. It is not only the case that queues for awaiting one's turn to mark attendance are always separate for men and women. Very often, male workers usually randomly stand, while the women workers are subject to stringent rules to stand in a queue and await their turn. Even if the women workers are made to stand in line first, male workers standing around haphazardly are permitted to mark their attendance first and leave. Hence, the culture of discipline and subsequent reprimand in case of any issues arising is always higher for women than male pourakarmikas.

And as a final blow to the women workers who were struggling, along with the BBMP, to get rid of the ghost workers who were siphoning off their wages, many contractors were successful in penetrating even the biometric system with phantom workers. The means by which this was materialised was ingenious. Workers reported that they could see spouses, sisters, friends, as well as other female acquaintances fraudulently registering as pourakarmika workers (providing all relevant identification), casually arriving every day at biometric attendance spots at the given times, departing at once from the spot after providing their thumb impression, and going about their ways for the rest of the day. A good part of the salary received in their name was shared with their respective contractor, in an enterprising nexus to deceive the BBMP and the public. Also, due to the well-acknowledged collusion between some staff of the BBMP and the contractors (as well as due to the political might of the contractors' own union), only a very small number of individuals are believed to have "won" the tender to supply the biometric scanning devices, which simplified the opportunity to tamper with the devices themselves.[11]

Hence, women pourakarmika workers, in general across the city, now face equally confounding work experiences even in the new technological regime – and not incidentally.

Beginning to unearth beneath the technological experiences

The experiences around the new technology constitute issues of exclusion, work discipline, insecurity within work conditions, experience differentials based on gender, and fraud. Drawing from the lessons of Cooper's (1988) study of the cigar industry, we learn that a technical study of the cigar machines may be only skin deep, but when examined within historical, labour, and gender contexts, there is much to learn. As we can see here, too, the constituents of technological experience are far beyond merely the technological – in fact, apart from issues of connectivity, there are few other really "technical" issues with the biometric attendance system – if we zoom out. A close reading of these constituents summons us to unearth what lies underneath the simply technological. We are required to unpack the mechanisms at the core of the experiences around the new technology. We are called upon to verify whether the technological intervention was based on a shallow understanding of socio-technical outcomes. And most of all, we are compelled to study, underneath the exterior, how informality, caste, and gender stand at the root of technological experience.

To begin this unearthing process, we need to place the experiences around the new technology within the cauldron of the larger work experiences of women pourakarmikas. The purpose of this is not to romanticise the examination of the technological experiences but to provide greater analytical rigour by studying them in their true context – in the everyday work experiences. While this exercise, in the section that follows, might appear to digress from an understanding of technological experiences, the conceptual analyses that follow a narration of the work experiences weaves the tapestry in its entirety.

Everyday work experiences

After recruitment and assignment of wards and streets to the pourakarmikas and associated workers, the contractors are required by very clear and detailed regulation (see BBMP, 2017a, 2017b) to provide these workers with a minimum level of safety and hygiene-related equipment. However, it was observed, and confirmed from speaking to the workers, that besides jackets (which serve mainly as visible identifiers on the street), little else is provided to the large majority of workers. Workers bring their own footwear and most of the time operate with bare hands and arms. Mouths and noses are covered with cloth that the workers bring on their own; ears are rarely covered at all. While brooms are arranged by the workers too, the scraping and collecting of solid waste on the streets is done either by hand or by cardboard sheets that the workers source on their own. There is blatantly recognisable gender segregation of work on the streets in the actual manual collection of solid waste. Male workers at this stratum do not usually collect the actual garbage from homes and streets (apart from some exceptions). While women workers are generally bending, sweeping, and manually collecting the solid waste, male workers either follow these women in auto-rickshaws, or at maximum, help transfer collected waste by avoiding its handling. The sight of male workers physically removing the garbage is very rare, and usually occurs in unpopulated areas or only when there are no women workers accompanying them.

Given the physical strain of the job, using toilet facilities becomes a regular requirement, especially for women. In these circumstances, women workers simply avoid urinating or defecating, and bear severe duress for their menstrual needs. Male workers, on the other hand, take advantage of the socially accepted habit of men resorting to open urination. It was also gathered from interviews that workers are simply at the mercy of either sympathetic households for drinking water, or have to seek washing sources at a distance. Food is regularly eaten amidst garbage, which is often swarming with flies and maggots; this makes them resort to purchasing meals at local eateries, adding to financial stress.[12]

Women workers are at a higher risk of losing a day's pay, not only due to their own occasional medical conditions but greatly due to their status at home as primary (or sole) caregivers to ailing elderly parents or sick children. The care of pregnant or nursing daughters or sisters, or obligatory demands from the extended family, regularly force women workers to take off from work, with no provision for earned or sick leave. Maternity leave is entirely nonexistent, which emerges as a serious issue because most women workers fall within childbearing age. Hence, in a given month, there is a high chance of at least a few days' leave from work, with the odds of this being much higher for women than men. This adds to the general problem in this occupation that they are subject to a seven-day work week, enjoy no listed holidays, and certainly no paid leave (besides namesake half days on Wednesday and Sunday).

On regular exposure to garbage – wet, dry, and toxic – skin ailments are considered routine and internalised as a part of their job. The risk broadens from skin diseases to malaria, tuberculosis, septicaemia, a range of gastric and respiratory ailments, and regular eye infections. In this foul work environment, on the streets or at the waste collection centre, menstrual hygiene for women is also severely compromised, leading to infections of the reproductive system. Access to regular health check-ups is nonexistent from the part of the state or contractors, and avoided by the individual workers out of cost concerns. Many male workers resort to alcohol consumption or heavy smoking to help withstand the stench, while women workers, facing lower social tolerance towards alcohol consumption, take recourse to chewing tobacco. These strategies, already fraught with danger on a daily basis, lead to an increased risk of developing cancer over the long term.[13] Hence, unhygienic conditions of work, coupled with negligible access to healthcare, have forced them to internalise the medical risks of their work.

Enquiries during fieldwork elicited a unanimous response on the rampant and habitual verbal abuse from supervisors, health inspectors, residents, passers-by, and several other such actors in their ambience – so much so that it has become a familiar and established state of affairs in their line of work. Especially in the case of women workers, working within a society which has been conditioned into believing that the job of cleaning by default is a feminine endeavour, the tone of voice from residents (even from domestic workers within those households) was reported to be typically harsh and degrading. This was regardless of the economic affluence of the neighbourhood. It was reported by our respondents that on one occasion, residents even clicked photographs of pourakarmikas resting and chatting after a day's work, and sent them to the BBMP as evidence of lethargy at work, for which they were severely reprimanded. Even if verbal abuse from residents may be occasional in some areas, what is regularly faced – in an even more dehumanising tone of voice and choice of vocabulary – is verbal abuse from the part of the supervisors and contractors. Again, by virtue of being women, informal workers, and being of Dalit caste as well, a certain sense of natural subordination is expected by contractors, by which they bestow on themselves the liberty to address the women pourakarmikas in as degenerate a manner in intonation and spirit as possible.

According to our respondents, as much as they are subject to this vile treatment, pourakarmikas are also able to influence the site of work based on their relationship and rapport with their supervisors. A placating and subservient demeanour can pave the way to being placed at a relatively cleaner location that is convenient to transport, food, toilet facilities, and the biometric attendance marking point. On the other hand, our respondents related how one woman worker who argued with her supervisor was penalised by being arbitrarily posted at a street around a kilometre away from these conveniences. She was subject to walking that distance each day with her loaded trolley, and compelled to rush between the biometric attendance point and her worksite several times a day.

Analysing these work conditions

Subsequent to the narration of everyday work experiences, it is inevitable to analyse these experiences and conditions as the penultimate step to the understanding of technological experiences. The work conditions of pourakarmikas, which have been insufficiently provided a rigorous academic treatment,[14] are more effectively analysed by placing *gender* as the central category of observation and analysis.

Gender as a principal analytical component in the critical analysis of the politics of labour and global capitalism has been well established in the academic literature.[15] Gender not only illuminates inequalities within power structures and power equations in specific spaces but as an analytical centrepiece escorts us towards understanding how these inequalities are extended into nearly all domains or spaces of everyday work and social life (Scott, 1985). Multiple and simultaneous conditions create and sustain these gendered inequalities in spaces of economic production or service-provision (Prakash, 2015). The workplace, hence, emerges as a special case of cultural reproduction, besides capitalist production (Chari and Gidwani, 2005), of the gendered environment in all walks of life and livelihood in contemporary economy and society.

A critical observation of the contemporary labour scenario in developing economy settings reveals how women workers have generally ended up in informal employment and informal survival strategies, within workplaces and occupations which harbour supposedly "natural" domestic responsibilities and "feminine" tasks. Gendered ideologies support flexible modes of labour control and discipline because of their ability to *naturalise* claims about the "worthiness" of the occupation, and what kinds of bodies are "suited" to particular tasks (Mills, 2003).

As has been observed in the context of home-based work, employers are known to prefer middle-aged mothers, as these workers are perceived as naturally self-disciplined, responsible, serious, and punctual (Boris and Prügl, 1996) – which also happen to be relatively more "feminine" values that are suited for the kinds of tasks related to home-based work. In fact, in some work settings these natural and feminine attributes are considered "useful" and a win-win outcome for both potential employers and women themselves, as the latter may otherwise "not find work" (Mullings, 2004). In factory work, for instance, women workers and their families in some cases voiced some "gratefulness" for these jobs, for if not for this, they might have been relegated to low-paid rural or agrarian employment, despite being school educated (Padmanabhan, 2012). If for women workers these jobs "save" them, for employers a female workforce implies reduced costs, lower probabilities for unionisation, easier recourse to compromising on decent work and safety, and little need for social security (Harriss-White, 2010; Wright, 1999, 2001). Even in local economic settings, local capital is known to use its deep knowledge of local cultural norms about women's docile roles in domestic spaces, which they are expected to reproduce at the workspace as well (Chari, 2004; Mills, 2003; Neethi, 2012; Padmanabhan, 2012). Women workers' skills (as well as challenges such as body aches or fatigue) are simply taken for granted, as they are considered an expected extension of domestic work, or as a cultural value transmitted naturally and informally by women through female progeny (Celia, 2010; Elson and Pearson, 1981a). Hence, while feminised activities carry biological determinism as an underlying principle, it is *cultural* determinism that is strategically used to establish it as the rational choice for women's work outside the home as well (Antony and Gayathri, 2008). Even with regard to the separation of home and work, while male workers are permitted to choose that separation to their advantage, for women workers this separation is blurred (often nonexistent), as they are expected to carry "human preoccupations" to the job and constantly bear in mind the larger social purpose of work (Cockburn, 1983). This larger social purpose or social value of work is cast as inseparable from the subjective meanings ascribed to women's work, which derive from local cultural histories and class structures, in turn closely linked to characteristics of power, class, and social identity (Upadhya, 2016).

The naturalisation and feminisation of jobs, worksites, and occupations solidify the already robust and historically established link between professional attitudes and traditional female roles (Neethi, 2014). This is the reason why labour markets for certain occupations – food processing, chemicals, rubber, plastics, plantations, besides garment work – target a specific gender, besides specific ages and ethnicities (Momsen, 2010). Jobs that are more time-intensive and dextrous are more likely to be feminised (Antony and Gayathri, 2008). And as an extension, work settings such as garment industries have preferred to feminise their workforce to benefit from a set of social relations in production which is wholly re-carved on the basis of gender, and which seamlessly reproduces domestic labour systems and patriarchal practices (Harvey and Scott, 1989; Harvey, 1991; Chari, 2004).

Bringing this entire analysis to our pourakarmikas facilitates us to grasp the fact that their everyday labour experiences are neither incidental nor specific only to them but are an aggrandised version of general trends in feminisation of workforce and ascription of domestic and social cultural values to women's work as well. Culturally, the division of labour in domestic spaces automatically ascribes the task of cleaning and waste management within the household and family as feminine labour. In a neat extension, the physical job of street cleaning and subservience in the work hierarchy of solid waste management automatically becomes rationally and naturally feminine, to be ascribed as women's work. At this point in the analysis we may propose an interesting extension to the concept of "nimble fingers" by Elson and Pearson (1981b), which argues that women workers are generally ascribed to certain occupations out of their natural capabilities to handle tasks that are cast as socially appropriate for women. In our analysis of the pourakarmikas, if their work on the streets is a natural extension of the skills and capabilities to handle cleaning tasks at home from early on in life, do they then possess the virtue of *coarse fingers* suited to pourakarmika work? In fact, this may be one of the reasons why pourakarmikas are rarely provided with safety equipment – after all, if it is absurd for women to require safety equipment while cleaning and cooking at home, it ought to be unnecessary on the streets, too.

Yet another naturalisation that rears its ugly head is that of caste-based division of labour, which has immediate connection with solid waste management. As mentioned earlier, the job of manual scavenging, street cleaning, or other hygiene-related manual labour has been historically populated by Dalit castes out of traditional occupational ascription, often even out of coercion and intimidation. This has continued at large even in the contemporary neoliberal era. Drawing from the work of Harriss-White et al. (2014), caste and religion have not eroded but instead have conveniently been embedded into the working of the state and economy, where forces within state and market have intensified social forms of regulation or have even created new forms. Even in urban metropolitan settings, this embeddedness is neither uncommon nor unsurprising, as subaltern groups – which here can include women of the underclasses, especially if they are of Dalit caste – are left to negotiate informal and casual labour markets, given that the more formal venues of employment in urban economies (for both men and women) require a certain stock of social and economic capital for entry and sustenance (Chalam, 2011; Das, 2010). It was reported by several pourakarmikas that while they had fled their rural origins and come to the city to escape caste and unemployment, they had ended up jumping from one whirlpool to another, as here, too, contractors were of the same dominant caste back in their villages, and the oppression now simply took on an urbane flavour. So as we can see, the conception of urban spaces as conducive to a caste-free social milieu does not harmonise well with lived realities in the city where caste continues to be a major source of oppression, and where occupation and caste (and connotations of untouchability) are still tightly bound (Chitageri, 2004; Gooptu, 1993). As seen also in the case of New Delhi (Prashad, 2000), rural lower-caste communities, upon migration to urban settings, are susceptible to immediately be visualised and automatically employed as "sweepers" at their urban destination.

What is revealed here is that women pourakarmikas stand at the crossroads of grave disadvantage, where gender stands out significantly as one of the axes of this intersection. While they belong to Dalit castes and are urban informal workers, most importantly, they also happen to be *female* workers. Male workers – including male pourakarmikas, auto-rickshaw drivers and helpers – on the other hand, have a relatively diminished experience in this regard. They do suffer the travails of belonging to Dalit castes and being pushed into informal labour settings. But by virtue of their gender, they are not subject to extrapolations – of "natural" attributes, "feminine" work-discipline, or "traditional" domestic values – in the workspace. Male workers also suffer from the vagaries of urban labour informality and have, in the first place, opted for this occupation either as distress migrants from rural hinterlands or because there is little way out for aspiring individuals within Dalit communities to gain employment (as seen in earlier chapters of this book). However, they are rarely the subject of verbal or sexual abuse, have much lesser stigma around sanitary practices, and are rarely required for domestic caregiving responsibilities (and, unlike women pourakarmikas in our case, probably do not participate in these chores before leaving home or after returning). Most of all, they are seen to help out their own female compatriots in the manual job of handling the garbage only in exceptional circumstances (as detailed in the earlier section). They are not "expected" to do the actual manual cleaning and collecting job if there are women workers around. This was reported by male pourakarmikas themselves, who stated that they were uncomfortable sweeping streets in populated areas, as this wasn't what they were "supposed to do"; with the contractors even mocking them, suggesting that they engage in work that is above this lowly feminine work, and often placing them in less populated streets. While there is absolutely no guideline in recruitment within this sector that women workers are preferred for pourakarmika tasks while men are preferred for driving and helping, Tables 6.2 and 6.3 display the realised outcome of processes that have operated at the intersection of informality, caste and, most critically, gender.

It should then be recognisable why (a) two-thirds of pourakarmikas are women, (b) nearly all drivers and helpers (and of course, supervisors) are men, (c) male workers in the sectors are neatly divided into pourakarmikas and non-pourakarmikas, and (d) most important of all, nearly *all* female workers in this sector are engaged as pourakarmikas.

Borrowing from the analysis of Beneria and Roldan (1987) and Hsiung (1996) in other economic settings, our pourakarmika workers appear to be standing at the intersection between the logic of capitalism (in terms of its preference for informal labour), the logic of caste (hailing from Dalit communities), and the logic of gender itself (beginning from domestic and social spaces, and reproducing to feminisation of workspaces). They are therefore situated within a "matrix of domination" (Prakash, 2015) structured around informality, caste, and gender. Borrowing from the work of Rege (1995, 2000), these women are afflicted with the triple oppression of caste, the sexual division of labour, and division of sexual labour.

And it is upon this sturdy analytical platform that we unpack the gendered experiences of the biometric attendance system.

Unpacking the gendered experiences

A rereading of the everyday encounters around the biometric system would now appear quite evidently coloured with a tone that highlights a deeply gendered experience. To commence our interpretation, we recognise the fundamental fact that the entry of a new technology in a production or service system is not a neutral but essentially a socio-political process (Schwartz-Cowan, 1983) – a fact that is now widely accepted in the literature on technological change and intervention as a cardinal principle. Schwartz-Cowan (1983) continues this line of thought, prodding us to undertake a sociological analysis by analysing not only the process of technological change (wherein an artefact replaces an earlier technological regime) but also social structures and diffusion through its designated space. These propositions have become rooted in the established and growing stream of literature on the feminist politics of technology, which has successfully permeated mainstream thought in the West on technological change, influencing interpretations of technological processes to eschew the notion that technology is the product of streamlined and linear rational technical imperatives that are sociologically neutral. The literature on feminist evaluations of technology has complicated the interpretation of technological experience with substantial historical and contemporary empirical evidence that technological outcomes and gender interests/identities are mutually shaped, for the most part in a convoluted, complicated manner (Wajcman, 2010). New technologies are invariably coupled with the established and institutionalised patterns of power in society, hence their use (or misuse) are indubitably embodied of gendered power relations in all spaces, including workspaces (Wajcman, 2004). Understanding what underlies the everyday experiences of women with new technology requires an appreciation of the fact that the specially disadvantaged outcomes for women are a natural result of the social embeddedness of technology (Sassen, 2002). Drawing further from Sassen (2002) and Kamath (2018) to interpret these experiences, clearly, it can be seen that social variables such as power, inequality, hierarchy, and so on, bring out variability in technological experience. With this approach, we must depart from a purely technological interpretation of the artefact in question, and read our everyday technological experiences in the context of the outcomes they have for particular social cohorts (Sassen, 2002). Even from the point of introduction, let alone their deployment, innovations are not divorced from the larger social and cultural world. In this case, the biometric system entered the workplace riding on the dismal concoction of informality, caste, and gender.

Returning to a theme we have already invoked earlier in this study, the nature of supervision of women within the domestic sphere, which is usually intimidatory, is also mirrored in the workspace by male-populating the supervision of women workers. We have recognised that the crudities of male supervision – not least intimidation – are aggravated in a workspace that involves the physically exerting an hygienically uninviting job of garbage collection and street cleaning. However, it is presupposed that the nature of supervision in the task of *attendance* at the workplace would be gender-neutral subsequent to the introduction of a new technology – by default assumed by the intervening body (the BBMP and Government of Karnataka) as socioculturally neutral. This is a careless extension of the popular perception that such technology is gender-neutral and, ambitiously, a "non-misusable" weapon to raze down institutionalised practices and power equations shaped by patriarchy and maybe to even "liberate" women (Wajcman, 2004). The artefact *on its own*, of course, does not differentiate the finger imprints of male and female workers. However, on the contrary, the new technology here has served as a sophisticated vehicle to facilitate the accurate reproduction of the intimidatory supervision, instead of producing a flat and politically contour-less outcome. This adds to the argument that technologies that enter such workspaces in the task of supervision, enter embodied with masculinity by their very construction (Cockburn, 1981; Wajcman, 2004, 2010; Corneliussen, 2012).

The operation of the artefact on the ground has been swayed by strongly gendered work and supervisory practices in this occupation, i.e., the actual implementation of this technological intervention has been modelled around established gendered practices (Lerman et al., 2003). Recall that the introduction of this technology had an entirely different purpose – to divorce the *pourakarmikas'* payment system from the contractors' dubieties, and to avoid ghost workers. We assume that not only were expectations of dismal everyday experiences overlooked from this goal (as is often the case with ICT-led interventions) but that even gender-neutrality as an assumption itself was entirely missing in deliberations at the BBMP around the introduction of this intervention. This reflects an imperceptibly deep gender-neutral technological determinism at the policy level. Clearly, therefore, there has been a mismatch between its purpose as imagined and the experience in reality. As in an earlier chapter in this book, we recall Akrich (1997), who argues that when unsuccessful outcomes arise, designers have sometimes tended to accuse users of having "misused" a device, while the reality probably tends towards a shallow understanding of the overlap of technology and gender by not only designers but especially policymakers. The technological experiences around a seemingly gender-neutral device such as an electronic attendance marker is the result of an intervention operating squarely inside the established crossroads of informality, caste, and gender. The peculiarities of the experiences around the biometric attendance system are, drawing from Wajcman (2004), both a source and consequence of entrenched gender relations in society and the sector. We may then even question what technology-based "improvement" at the workplace actually is – does it mean progress or empowerment only among the privileged in the gender equation, given that technologies are never gender-neutral and rarely have "general consequences" (McGaw, 1996; Corneliussen, 2012)?

Conclusion

Pourakarmikas – a subaltern informal feminised workforce in an urban setting (with coarse fingers) – provide a living example of the everyday experience of these realities. With this study we have first, empirically, supplemented the thin academic documentation of the travails of pourakarmikas in Bangalore. Second, analytically, we have placed gender as the centrepiece along with caste and informality in the examination of their everyday work conditions and technological experience. Third, conceptually, we provide a combined analysis of informality and socio-technology in their regard.

It ought to be no surprise that everyday work and technological experiences are outcomes of an amalgam of multiple social variables that have conjointly built the deeply downgraded footing for these female informal workers. And overall, upon adopting this analysis, it ought to also be of little surprise that the pourakarmika workforce is predominantly female, and the female workforce is nearly entirely involved in pourakarmika tasks. Technological experience here has reflected and reproduced informal-labour experiences. In other words, in the case of the pourakarmikas in Bangalore, the new technological intervention has simply seated itself in the existing sociopolitical ambience. That is, the new technological experience here has been not about straightforward logics but essentially about reproducing sociocultural power (Corneliussen, 2012; Lerman et al., 2003). This brings us to a seminal argument in the field of the politics of technology – that there are certain "un-engineerable" properties of technologies that are so because these technologies (and the spaces of their operation) are inevitably associated with particular institutionalised patterns of power and authority (Winner, 1980). To conclude, we assert that everyday work experience and everyday technological experience are founded on similar roots, and are hence neither incidental nor detached (or exogenous of one another) but actually *intertwined*. Both sets of experiences ride on the dismal concoction and resilient and thriving mechanics of informality, caste, and principally gender in this occupation.

Notes

1 Neethi P. is a faculty member associated with the theme Urban Employment at the Indian Institute for Human Settlements (IIHS), Bangalore, India. Saloni Mundra is Programme Executive at the Aajeevika Bureau, Udaipur, India.

2 A term from the Kannada language first employed by the government in 1972. See Parameshwara (2013) for an extensive treatment of the background of this term and its associated social groups.

3 We fully acknowledge that even this data with genuine contract workers might be plagued with errors with regard to the inclusion of ghost workers. However, the population of pourakarmikas as nearly 17,000 shown in this table can be approximated as the genuine figure.

4 Reported in *The Huffington Post* (20 October 2017), *The Times of India* (20 October 2017), and *The News Minute* (19 October 2017).

5 Reported in *YourStory* on 8 October 2017.

6 According to the pourakarmika unions and civil society organizations supporting them, this would have not been possible without some collusion with members of the BBMP.

7 Reported in *The Hindu* on 15 April 2017. Also reported here was that there were instances of wage differentials between male and female pourakarmikas, which were again the handiwork of contractors.

8 As stated by the Joint Commissioner of Solid Waste Management of the BBMP, reported in *The Hindu* on 23 March 2018.

9 While these issues were being slowly ironed out at the time of writing this study, the *Guttige Pourakarmika Sangha* was also pushing the BBMP and the state health and family welfare department to extend the *Jyoti Sanjeevini Scheme* (a form of universal free basic healthcare established in 2014 for state government staff and dependants for tertiary care) to pourakarmikas too, to abrogate health-related expenses and obstacles to regularity at work, as part of social security and risk aversion.

10 *The Report of the Committee on the Improvement of Living and Working Conditions of Sweepers and Scavengers*, April 1976.

11 As revealed during long interviews with civil society organisations. The possibility of tampering was reported by *The News Minute* on 9 June 2018.

12 The International Society for Krishna Consciousness (ISKCON), an international spiritual organization, collaborated with the BBMP to supply meals at 10 a.m. to over 30,000 pourakarmikas across 198 wards under its highly successful school lunch provision scheme *Akshaya Patra* (*The New Indian Express*, 14 April 2017). According to the ISKCON website, Rs.100 million was allocated by the BBMP in 2016–2017 to this collaboration not only to pay for the meals, but also to provide plates to eat them in. Unfortunately, issues of food quality (despite the BBMP paying for the meal) kept these workers away from this facility.

13 See Nagaraj et al. (2004) on the incidence of ailments among pourakarmikas, particularly women.

14 Discussions and the limited number of studies in the past have brought out valuable insights into the various aspects of pourakarmika life and livelihood. Chitageri (2004), though not exclusively on pourakarmikas in Bangalore, provides an interpretation on gendered caste relations in the context of urban space. Parameshwara (2013) and IIMB (2014) are also noteworthy. Other analyses have included Sangama (2002) on the work conditions in general, and discussion forums within institutions such as the National Law School of India (NLSIU), again on pressing issues around life and livelihood. The Centre for Labour Studies at the NLSIU held a roundtable on "Law, Poverty and Marginalisation; Reflections on the Life and Working Conditions of the Pourkarmikas in Bangalore" in June 2002. Also see "Caste Out Garbage Bengaluru" at

<casteoutgarbagebengaluru.wordpress.com> for a collection of other formal and informal writing on pourakarmikas.

15 See Neethi (2016) for a comprehensive treatment of this theme.

References

Akrich, M. (1997) 'The De-Scription of Technical Objects', in Bijker and Law (eds.) *Shaping Technology/Building Society: Studies in Socio-Technical Change*, The MIT Press, London and Cambridge, MA.

Antony, P., and Gayathri, V. (2008) 'Ricocheting Gender Equations: Women Workers in the Call Centre Industry', in Saith, Vijayabaskar, and Gayathri (eds.) *ICTs and Indian Social Change: Diffusion, Poverty, and Governance*, Sage, New Delhi.

BBMP (2017a) *Bengaluru's SWM Information Manual: Part 1: Overview*, Solid Waste Management, Bruhat Bengaluru Mahanagara Palike, Bangalore.

BBMP (2017b) *Bengaluru's SWM Information Manual: Part 2: Ward Specific Manual*, Solid Waste Management, Bruhat Bengaluru Mahanagara Palike, Bangalore.

Beneria, L., and Roldan, M. (1987) *The Crossroads of Class and Gender: Industrial Homework, Subcontracting, and Household Dynamics in Mexico City*, University of Chicago Press, Chicago.

Boris, E., and Prügl, E. (1996) *Homeworkers in Global Perspective: Invisible No More*, Routledge, London and New York.

Celia, M. (2010) 'We Are Workers Too! Organizing Home-Based Workers in the Global Economy', Women in Informal Employment: Globalizing and Organizing (WIEGO) Organizing Series, Bangalore.

Chalam, K.S. (2011) *Economic Reforms and Social Exclusion: Impact of Liberalisation on Marginalised Groups in India*, Sage, New Delhi, Thousand Oaks, London, and Singapore.

Chari, S. (2004) *Fraternal Capital: Peasant Workers, Self Made Men, and Globalization in Provincial India*, Permanent Black, New Delhi.

Chari, S., and Gidwani, V. (2005) 'Introduction: Grounds for a Spatial Ethnography of Labour', *Ethnography*, 6(3): 267–281.

Chitageri, S. (2004) *Uncovering Injustice: Towards a Dalit Feminist Politics in Bangalore*, Unpublished PhD dissertation, University of Warwick, UK.

Cockburn, C. (1981) 'The Material of Male Power', *Feminist Review*, 9: 41–58.

Cockburn, C. (1983) 'Caught in the Wheels', *Marxism Today*, 27: 16–20.

Cooper, P. (1988) 'What This Country Needs Is a Five Cent Cigar', *Technology and Culture*, 29(4): 779–807.

Corneliussen, H.G. (2012) *Gender-Technology Relations: Exploring Stability and Change*, Palgrave Macmillan, UK.

Das, M.B. (2010) 'Minority Status and Labour Market Outcomes: Does India Have Minority Enclaves?', in Thorat and Newman (eds.) *Blocked by Caste: Economic Discrimination in Modern India*, Oxford University Press, New Delhi.

Elson, D., and Pearson, R. (1981a) 'Subordination of Women and the Internationalization of Factory Production', in Young, Wolkowitz, and McCullagh (eds.) *Of Marriage and the Market: Women's Subordination in International Perspective*, CSE Books, London.

Elson, D., and Pearson, R. (1981b) '"Nimble Fingers Make Cheap Workers": An Analysis of Women's Employment in Third World Export Manufacturing', *Feminist Review*, 7(1): 87–107.

Gooptu, N. (1993) 'Caste and Labour: Untouchable Social Movements in Urban Uttar Pradesh in the Early Twentieth Century', in Peter (ed.) *Dalit Movements and Meanings of Labour in India*, Oxford University Press, New Delhi.

Harriss-White, B. (2010) 'Work and Wellbeing in Informal Economies: The Regulative Roles of Institutions of Identity and the State', *World Development*, 38(2): 170–183.

Harriss-White, B., Basile, E., Dixit, A., Joddar, P., Prakash, A., and Vidyarthee, K. (2014) *Dalits and Adivasis in India's Business Economy: Three Essays and an Atlas*, Three Essays Collective, India.

Harvey, D. (1991) 'Flexibility: Threat or Opportunity', *Socialist Review*, 21(1): 65.

Harvey, D., and Scott, A.J. (1989) 'The Practice of Human Geography: Theory and Empirical Specificity in the Transition from Fordism to Flexible Accumulation', in Macmillan (ed.), *Remodelling Geography*, Blackwell, Oxford.

Hsiung, P.C. (1996) *Living Rooms as Factories: Class, Gender, and the Satellite Factory System in Taiwan*, Temple University Press, Philadelphia.

IIMB (2014) *Study on Primary Collection of Solid Waste Management in Bangalore*, Centre of Excellence in Urban Governance, Indian Institute of Management, Bangalore.

Kamath, A. (2018) '"Untouchable" Cellphones? Old Caste Exclusions and New Digital Divides in Peri-Urban Bangalore', *Critical Asian Studies*, 50(3): 375–394.

Lerman, N.E., Oldenziel, R., and Mohun, A.P. (2003) *Gender & Technology*, The Johns Hopkins University Press, Baltimore and London.

McGaw, J.A. (1996) 'Reconceiving Technology: Why Feminine Technologies Matter', in Wright (ed.) *Gender and Archaeology*, University of Pennsylvania Press.

Mills, M.B. (2003) 'Gender and Inequality in the Global Labour Force', *Annual Review of Anthropology*, 32: 41–62.

Momsen, J. (2010) *Gender and Development*, Routledge, New York.

Mullings, B. (2004) 'Sides of the Same Coin? Coping and Resistance among Jamaican Data-Entry Operators', *Annals of the Association of American Geographers*, 89(2): 290–311.

Nagaraj, C., C. Shivaram, K.K. Jayanth, and M.N. Narasimha. (2004) 'A Study of Morbidity and Mortality Profile of Sweepers Working under Bangalore City Corporation', *Indian Journal of Occupational and Environmental Medicine*, 8(2): 1–18.

Neethi, P. (2012) 'Globalisation Lived Locally: Enquiries into Kerala's Local Labour Control Regimes', *Development and Change*, 43(6): 1239–1263.

Neethi, P. (2014) 'Home-Based Work and Issues of Gender and Space: A Case from Kerala', *Economic and Political Weekly*, 49(17): 88–96.

Neethi, P. (2016) *Globalization Lived Locally: A Labour Geography Perspective*, Oxford University Press, New Delhi.

Padmanabhan, N. (2012) 'Globalisation Lived Locally: A Labour Geography Perspective on Control, Conflict and Response among Workers in Kerala', *Antipode*, 44(3): 971–992.

Parameshwara, N. (2013) *The Role of BBMP in the Rehabilitation of Pourakarmikas in Karnataka: Special Reference to Bangalore City: A Sociological Study*, Unpublished PhD dissertation, Bangalore University, India.

Prakash, A. (2015) *Dalit Capital: State, Markets and Civil Society in Urban India*, Routledge, New Delhi.

Prashad, V. (2000) *Untouchable Freedom: A Social History of a Dalit Community*, Oxford University Press, New Delhi.

Rege, S. (1995) 'The Hegemonic Appropriation of Sexuality: The Case of the Lavani Performers of Maharashtra', *Contributions to Indian Sociology*, 29(1–2): 23–38.

Rege, S. (2000) '"Real Feminism" and Dalit Women: Scripts of Denial and Acquisition', *Economic and Political Weekly*, 5: 492–495.

Sangama (2002) *'Swaccha Bangalooru': Cleansed by the Sweat of the Poor*, Support Group for Contract Pourakarmikas.

Sassen, S. (2002) 'Towards a Sociology of Information Technology', *Current Sociology*, 50(3): 365–388.

Schwartz-Cowan, R. (1983) *More Work for Mother*, Basic Books, New York.

Scott, J.C. (1985) *Weapons of the Weak: Everyday Forms of Peasant Resistance*, Yale University Press, New Haven and London.

Upadhya, C. (2016) *Re-Engineering India: Work, Capital, and Class in an Offshore Economy*, Oxford University Press, New Delhi.

Wajcman, J. (2004) *Techno Feminism*, Polity Press, Cambridge, UK.

Wajcman, J. (2010) 'Feminist Theories of Technology', *Cambridge Journal of Economics*, 34(1): 143–152.

Winner, L. (1980) 'Do Artifacts Have Politics?', *Daedalus*, 109: 121–136.

Wright, M.W. (1999) 'The Dialectics of Still Life: Murder, Women and the Maquiladoras', *Public Culture*, 29: 453–473.

Wright, M.W. (2001) 'Desire and the Prosthetics of Supervision: A Case of Maquiladora Flexibility', *Cultural Anthropology*, 16: 354–373.

7

FINAL THOUGHTS

The Saliyars declined because they were too cohesive. The Dalit communities lost out on the mobile phone "revolution" because they were situated in a pernicious convergence. The pourakarmikas experienced a distressing engagement with the new biometric technology because they were informal women workers. In each case, the technological experiences and outcomes materialised on account of the underlying sociological structure and dynamics. Throughout these studies, I have urged for the necessity of observing, analysing, and assessing a techno-logical outcome as necessarily a sociological process, with the society–technology choreography being presented not as three exotic anecdotes but to demonstrate how society is an essential part of the reality of a technological experience. As stated right at the outset of this book, my goal is to exhort that there is nothing presumable or inevitable about a technological experience – even if that tech-nology is said to be endowed with far-reaching powers – rather it is contingent on the sociological crucible that the technology is sojourned in. I deliberately say *sojourned* in a sense of impermanence, due to the innateness of change in society, whether glacial or ephemeral. That is, this essential feature of society, namely its continual change, entails that technological experience – even with the same artefact or process – is also constantly subject to evolution and recasting, contingent on the speed and character of the sociological change. Thus, there is little that is either certain or linear about technological experience given that its marsupium – society – is itself subject to ceaseless moulding and mutation.

This is an awkward (maybe even vexing) stance to adopt in India, in a future we popularly imagine for this country that we uninhibitedly and so confidently call a "knowledge-based future" or "technological era." If we define a "technologi-cal era" as one where technology plays a significant role in life and work, human-ity has *always* been in a technological era. In fact, it can even be proposed that human societies form not only out of emerged commonalities in sociocultural elements but also out of shared technological elements – given the possibilities provided by nature to craft technologies, and the meanings society evolves and ascribes to the technologies it engages with and harbours. Societies and tech-nologies have always evolved together in our history, and therefore need to be analysed together. To understand the one, we have to study the other. When one metamorphoses, so does the other. If certain attributes within society are inelastic, technological experience associated with that society will reflect those uncompromising attributes. For this reason, we must not fall to anticipating that technological change will assuredly pilot modernity, whether as temples in the form of big dams or as compact sleek panaceas such as mobile phones. National plans and policies, media and culture, specious visionaries, the economy – all these usual and unusual suspects must be restrained when in the process of being smitten in admiration towards the ascribed mythical virtues and powers of technology, in its almost patronising shepherding of society's trajectories.

To reiterate what I had proposed at the beginning of this book: there is an urgent need to rehabilitate the fissures between sociological and technological understandings in India, by promoting the acknowledgement, appreciation, and inclusion of the social context within the technological experience. I have especially attempted to reveal how hollow our policy understanding is on the historical and continuing forces of society–technology interaction, and how popular perceptions have naively disembodied this linkage. To appreciate the complexity of these relationships and to improve the porosity between these two spheres, I have employed a variety of concepts, methods, sites, agents, and subjects, by which I have also sought to lay bare the cavernous void in the breadth of our understanding of our development challenges, both today and in the future.

In development policy, technological-determinist tendencies must be watched out for, and terms such as "information society" must be employed with careful consideration. In development action too, it must be recognised that because tangled sociopolitical factors can generate unexpected technological outcomes, technological interventions to address quandaries on the ground must be deployed with an intricate understanding of the society–technology nexus that thrives in the locale. Technological intercessions have done wonders on a great many occasions around the world, but the enchantment around these triumphs must not encourage complacency, whereby convenient and overconfident extrapolations of these successes are drawn across the board. We must not give in to hagiographic accounts of technological outcomes, even if genuinely successful, as technological experiences are intensely participatory processes (even if they are not always democratic or equitable). Technological experience can be heterogeneous for the same innovation across social groups, and no outcome is "inevitable." Technological intervention must be reflected upon, ex ante, as a sociologically inclusive process, as social structures and dynamics enter at every stage of technological experience, as they do for nearly all economic relationships.

Not just for weavers or historically depressed communities or garbage sweepers – every individual, in their relationship with technology, is actually wading through a socio-technological experience while engaging with that technology. In my previous volume *Industrial Innovation, Networks, and Economic Development*, I proposed for more serious *sociological* enquiry on innovation and learning processes in India, in a departure from the purely economic or technical foundations of mainstream discourses around these processes. I advanced this need for greater enquiry into the economic-sociology of technological processes, because as much as they are economic interests, technological outcomes are pivoted around sociological interests too, and are sociologically conditioned. This is an imperative revival or awakening that is necessary for technological analyses in India, given that each facet of the sociological labyrinth that India is, has its own socio-technological experience that attracts serious perusal. I attempt and repeat the call for these pressing concerns in this volume too, but on this occasion, I have taken the opportunity to append this call for a vital task that is characterised by a gaping vacuity. Through the subjects studied across the chapters in this book, I have implored for further enquiry and writing not only on the social context of technological experiences but also on the technological experiences of the *subaltern* in India. While there is an immense and growing body of academic and popular literature on the myriad pressing concerns among the subaltern, the literature on the technological engagement of the subaltern is still in want. The current body of work is thin and desperately needs attention. The reason for this clarion call is twofold. First, if we do envision our future as "technological" (whatever that might mean), and if our technological experiences are sociologically conditioned, then the subaltern concern around technological experience warrants an exclusive analysis. Second, given the triumphant advancement of neoliberalism permeating literally every facet of life and work and society, exclusionary and non-democratic tendencies in the functioning and reconfiguring of economy and society are gaining robustness, implying that the interests of the subaltern will only be sidelined, especially technological interests. There is scant writing and documentation of the subaltern experience of technology, which this book makes a concerted attempt at.

So much more is yet to be done. We need to reconfigure our understanding and expectations around technological change, as it does not have an autarchic existence or autonomous operation outside of society. We mould ourselves around the technologies we create, but much more so, our technologies are what we are.

INDEX

Note: Page numbers in **bold** indicate tables and page numbers followed by 'n' refer to notes.